Brent & Jane,

With best wishes
to you, your family or
your business. Thank,
your wisdom, advice and [?]

Craig

The Business of Bioscience

Craig D. Shimasaki, PhD, MBA

The Business of Bioscience

What Goes into Making a Biotechnology Product

 Springer

Craig D. Shimasaki
BioSource Consulting
InterGenetics Incorporated
14200 S. MacArthur Blvd.
Edmond, OK 73025
USA

ISBN 978-1-4419-0063-0   e-ISBN 978-1-4419-0064-7
DOI 10.1007/978-1-4419-0064-7
Springer Dordrecht Heidelberg London New York

Library of Congress Control Number: 2009927502

Printed on acid-free paper

Springer is part of Springer Science+Business Media (www.springer.com)

*To Verna, my loving wife and friend of
31 years, who has always stood by me with
support and encouragement. To Alyssa and
Lori, my heartwarming daughters, who by
their very presence always inspired me to
be a better father. To my parents, Jack and
Betty, who allowed me to be everything
I could ever hope to be, and always
supported me unconditionally. To God,
who constantly inspires me with insight
and vision, and makes all things possible.*

Preface

My journey into this fascinating field of biotechnology started about 26 years ago at a small biotechnology company in South San Francisco called Genentech. I was very fortunate to work for the company that begat the biotech industry during its formative years. This experience established a solid foundation from which I could grow in both the science and business of biotechnology. After my fourth year of working on Oyster Point Boulevard, a close friend and colleague left Genentech to join a start-up biotechnology company. Later, he approached me to leave and join him in of all places – Oklahoma. He persisted for at least a year before I seriously considered his proposal. After listening to their plans, the opportunity suddenly became more and more intriguing. Finally, I took the plunge and joined this entrepreneurial team in cofounding and growing a start-up biotechnology company.

Making that fateful decision to leave the security of a larger company was extremely difficult, but it turned out to be the beginning of an entrepreneurial career that forever changed how I viewed the biotechnology industry. Since that time, I have been fortunate to have cofounded two other biotechnology companies and even participated in taking one of them public. During my career in these start-ups, I held a variety of positions, from directing the science, operations, regulatory, and marketing components, to subsequently becoming CEO. During the past two and a half decades, I have learned more than I thought was possible about starting, building, and operating a biotechnology company. I experienced the joys and disappointments that come with this unique merger of science and business. Biotech entrepreneurs experience firsthand the transformation of a scientific idea into a real product, the challenge of building a company infrastructure, issues in dealing with regulatory requirements, manufacturing, and market issues. They learn how to overcome challenges in such a way that is both satisfying and rewarding. Until you have been a part of a biotech start-up, you will never know the thrill that comes from creating and bringing to market a life-saving product, conceived and developed in part, from your own ideas, and the work of your hands, and those of a talented team. Surprisingly, the most important lessons are usually learned, not from successes, but from apparent "failures," as you work through them to transform these situations into something better than was expected or anticipated.

When I first journeyed down this career path, I had no formal guidance or back-ground in being a biotechnology entrepreneur. Although I was prepared for scientific rigor through my PhD training, then operationally through my MBA qualification, I did not appreciate the breadth and complexity of these issues until I faced these challenges on a daily basis. Consequently, I had to learn on-the-job while shouldering the scientific and the business needs of a company.

Biotechnology Industry Limitations

A severe limitation to the expansion and maturation of this industry is the availability of seasoned entrepreneurs and team members, who understand what it takes to establish and grow successful life science companies. Most of the training for biotech entrepreneurs comes from firsthand experiences while on-the-job. This learn-as-you-go method is not an ideal way to ensure success of these organizations because the product development process is complex, and the financial stakes are extraordinarily high. Several universities and business schools now offer programs specializing in training biotech leaders and entrepreneurs – yet the availability of experienced life science entrepreneurs is still in significant deficit. In the US, numerous states have biotechnology initiatives supported by government funding to foster the creation of biotech and life science companies. Sadly, however, all admit a dearth of seasoned biotech leaders to shepherd these fledgling organizations. To fill this void, most organizations support an "Executive-in-Residence" or "Entrepreneur-in-Residence" program, hoping to find experienced biotech leaders to spawn and nurture development of these new biotechnology companies. These programs plan to recruit and relocate seasoned biotechnology leaders from other states to lead their newly developed biotech companies. Unfortunately, each organization is vying for the same limited resource of individuals, with no state having an excess of experienced biotech leaders to keep up with the demand in this industry.

Who This Book Is for?

The Business of Bioscience: What Goes into Making a Biotechnology Product is for all those fascinated by the biotechnology industry, and those who want to know of the challenges and rewards that accompany the development of a biotechnology product. It is for those interested in beginning, or joining, a development-stage biotechnology company, such as the entrepreneur, or would-be entrepreneur, and interested team members. It is for the critical support groups, those whose work and efforts support and partner with biotechnology, such as patent attorneys, executive recruiters, and regulators to name a few. This information will be of interest to those who finance this industry, such as venture capitalists and angel investors. This book was also written for other professionals who have an interest in this industry.

You may be an investor, analyst, or someone yearning to learn what goes on behind the scenes in the building of a biotechnology company. No matter who the reader is, you will learn things applicable to help work effectively with, and in the biotechnology industry.

Those who seek guidance and practical insight into the challenges that lie ahead in building a biotechnology company will find this information invaluable. You may be a former pharmaceutical executive with many years of experience; you may be a professor, scientist, or physician with the gnawing desire to build a business based on a breakthrough medical technology. This book can be a road map for the life science entrepreneur to help navigate the labyrinth of starting and building a successful biotech company. Even experienced biotech professionals will gain additional insight by seeing the world from the eyes of an entrepreneur, to better understand what they can do to participate in the growth of their biotech companies. Finally, for those who just want to know the inner workings, challenges, and opportunities within such an organization, they will find that the process of developing a biotech product is not as straightforward as one may think.

It is my hope that *The Business of Bioscience* will also serve as a teaching guide and companion for business schools and universities with programs focusing on the entrepreneur in the biotechnology industry. Most business schools today have an entrepreneurial focus, and some are developing programs specifically supporting the life science industry with its unique set of issues and needs. This book is intended to serve as a resource for institutions teaching a growing number of students in the emerging, and life-changing field of biotechnology entrepreneurship.

Why I Wrote *Business of Bioscience*

During my years as a formative biotech entrepreneur, I painfully recognized a void of knowledge within myself about how to establish and build a biotechnology company. To fill this void, I voraciously read dozens of books on various topics that I believed would help me as a biotech entrepreneur. These books covered topics on general business development, basic marketing, project management, writing business plans, employee hiring practices, leadership principles, high-tech company histories, corporate alliances, fundraising, regulatory affairs, in addition to my regular scientific journals and biotech industry publications. Although each of these self-help books contained a component that could be applied to my entrepreneurial interests, none spoke directly to my situation as a biotech entrepreneur. Portions of these books contained bits of information that were generally applicable to all industries. These topics included: "how to write a winning business plan," "how to hire good people," "raising money," and other general entrepreneurial topics. But like a wrong-sized glove, nothing exactly fit my situation. What was missing from these books was the integration of these topics, within the context of the biotechnology industry, with examples of real-life challenges the biotech entrepreneur faces on a daily basis.

Over the years, my desire was to summarize my learning experiences, and to write a book that was encompassing and relevant to equip future biotech entrepreneurs. I wanted this book to be practical, with insight that could be applied to various stages of biotech business development. In addition, I wanted this book to be enjoyable to read, not for entertainment sake, but to help the reader remember the information and apply it to the right situation.

What Is Covered in This Book

In *The Business of Bioscience* you will be introduced to many key aspects of the biotech and product development process. This includes: the characteristics of a biotech entrepreneur, how to establish an organization, raising many forms of capital, selecting development milestones, dealing with personnel issues, the significance and impact of a business model, market strategies and their impact on raising money, regulatory compliance issues, and why you must manage differently at different life-stages of the organization. I have included background information at the beginning of each chapter so as not to take for granted the readers' understanding of all subject matter; yet we quickly move to in-depth discussions of these subjects so that those with more experience in these areas will also benefit. Knowledge is essential for success, and most biotech entrepreneurs are very knowledgeable. However, a distinguishing difference between a good biotech leader and a great biotech leader is having wisdom – the ability to apply specific knowledge at the right time, in the right situation. This aspect is an overarching theme of this book.

As you read this volume, you will learn the mechanics and some business strategy of how to establish and grow a life science company. You will come away with an understanding of the expectations of biotech investors, and also gain insight on how to use this information to give you the best opportunity to raise the needed capital for a growing company. Securing funding for a biotechnology company is a crucial requirement if you hope to have any opportunity for business success. As a result, I have devoted two chapters to multiple aspects of the fundraising and financing process. Since there is literally not enough money available to fund all good ideas in this industry, it is important for you to understand the expectations of investors, and what constitutes a fundable company.

In this book, I touch on development life-stages in a biotech company and the parallel to life-stages of individuals. This analogy is helpful to understand how and when to transition a management style to meet the changing needs of an organization during company development. To be successful, you need information specific to the precise stage of growth of an organization. A company's needs have differing priorities based on the life-stage of the organization. For instance, you can read books on how to set-up proper financial accounting and compliance measures for Sarbanes-Oxley, which are always essential; however, if an organization does not have operating capital for more than 90 days, financial accounting compliances are not a high priority at this stage. Conversely, understanding how to write an effective business plan is important, but if the organization is having morale issues because of poor hiring decisions, it is time to address these before losing staff or seeing their motivation disappear.

The future of a start-up biotechnology company relies on decisions made at critical junctures. The consequences of some choices may seem subtle but can have a dramatic impact on the company. It is essential that the biotech entrepreneur have a working knowledge in all areas of company and product development, to make wise decisions, and ensure the future success of an organization. I have referenced additional materials on particular topics should you desire more in-depth knowledge on these subjects. In the future, provided that there is sufficient interest in this subject matter, it is my intent to add a workbook to support more formalized instruction suited to an academic course curriculum.

Throughout *The Business of Bioscience*, you will be learning technical information coupled with strategy, and placed into a context of the "bigger picture." This information is directly applicable to establishing and growing your organization, and developing your product and team, such that there is continuity in growth regardless of the size of your company. No matter how comprehensive a book, there will always be additional aspects of biotechnology entrepreneurship that can be covered; however, what is contained here, are the essentials that you will need to establish and grow your biotechnology company, and move your product development thorough to commercialization. The information in this book will help guide you through the rough spots and avoid a few detours along the way, and serve as a ready resource for help when you need it.

Summary

The biotech industry is in great need of seasoned biotechnology leaders who have the knowledge, wisdom, and desire to lead others to accomplish something greater then they themselves can do alone. This industry needs strong value-based leaders and teams of individuals who can see the future of medicine, and help usher the changes that genomics, proteomics, molecular biology, and personalized medicine can bring to human health. The overarching goal of this book is to equip, train, and prepare future leaders in the biotech industry, such that their products and services can advance medicine, that more may live longer, healthier lives. My desire has been to combine strategy, operational guidance, and practical wisdom in this book. My hope is that you will find this information valuable, as you join the ranks of entrepreneurs, visionaries, and scientific teams in establishing and building a biotechnology business. If I succeed in helping you avoid potential pitfalls in the entrepreneurial process, it will have been worthwhile. I am a firm believer in an ancient proverb, "learn from the mistakes of others, because you will never live long enough to make them all yourself."

I welcome any comments that you, the reader, may have on ways to improve this resource so that biotech leaders and entrepreneurs can be better equipped to lead our industry into the future. Additional information can be found on various topics helpful to the biotech entrepreneur at my website, www.businessofbioscience.com

Best Wishes for Your Success!

Craig Shimasaki, PhD, MBA
Edmond, OK

Acknowledgments

This book would not be possible without the many helpful individuals who read, reviewed, and proofed this manuscript and provided many helpful suggestions along the way. Thanks to Anil Gollahalli for his helpful review and comments and edits on intellectual property and university licensing. Thanks to Carl Gibson for his thoughts and helpful suggestions on the financing chapters. Thanks also to Tom Dickerson, who graciously shared information on the number of technologies reviewed and funded, and for his helpful advice over the many years. A special thanks to Anthony Hickey and Sean Ekins for taking the time to review portions of this manuscript and provide additional helpful insights and comments. Many thanks to my wife, Verna Shimasaki, for her helpful comments, edits, corrections and unwavering support throughout this process. Sincere appreciation goes to my new son-in-law, Adam Zarlengo, for the sharing of his experiences in a start-up company. A special thanks to the Gooden Group, the owner, Brent Gooden along with his business partner Jane Braden, who saw the value of what we were doing at two of the companies that I was a part of cofounding, and continued to stay with us throughout the tough times. A special thanks to Dale Hagaman of Accord Human Resources, whose company has supported our entrepreneurial work with their excellent services and with financial help. Grateful appreciation goes to the Founders and Board Members of the Presbyterian Health Foundation, who have always been supportive of biotechnology, and of our efforts in start-up companies. A very special thanks to Bill Swisher and his family. Bill has been one of the most ardent supporters of medical technology, and possesses an unwavering commitment to those he believes in, and is truly a man of his word. There are numerous supporters and friends that I would like to acknowledge and thank, who have stood by and supported the many company endeavors that I have been involved in over the years – but this would take an entire chapter. So I gratefully acknowledge them, as the "host of angels, supporters and colleagues," who tirelessly give of their time, and work behind the scenes.

Contents

Chapter 1
Introduction

Biotechnology is fast becoming one of the greatest scientific and medical revolutions in history. So far, according to the Biotechnology Industry Organization (BIO), biotechnology has produced more than 200 new treatments and vaccines that were incapable of being developed 20–30 years ago. Currently, there are more than 400 biotechnology drugs and vaccines in various stages of clinical testing, focused on treating more than 200 diseases such as heart disease, cancer, diabetes, Alzheimer's disease, multiple sclerosis, arthritis, and AIDS. The biotechnology industry has created hundreds of new medical diagnostics and laboratory-based tests, from molecular tests that detect the avian flu, to diagnostics that keep the blood supply safe from the AIDS virus. Because of this new industry, some of the most effective drugs in this century have been developed, drugs that work by mechanisms unimaginable a few decades ago.

It all started in 1976, with the discovery of the unique ability to engineer bacteria to produce human proteins of our choice. The first biotechnology products included human insulin, human-growth hormone, factor VIII, and tissue plasminogen activator (tPA). Now we have humanized monoclonal antibodies that specifically target septic shock proteins, B cells in non-Hodgkins lymphoma, and arthritis. Other humanized monoclonal antibodies target receptors for the treatment of breast cancer, and tumor necrosis factor (TNF) which treats various forms of arthritis. Biotechnology companies around the globe are producing tools and treatments based upon new discoveries that test and treat some of the most challenging medical conditions, so that more can live longer, healthier lives.

Biotech in Its Infancy

The biotechnology industry is a young and vibrant industry, that today, has a history of only 33 years. This is not so venerable when you compare it with the history of banking and finance with over 2,000 years, or the 230-year old

C.D. Shimasaki, *The Business of Bioscience: What Goes into Making a Biotechnology Product*, DOI 10.1007/978-1-4419-0064-7_1,
© 2009 American Association of Pharmaceutical Scientists

automobile industry – started with the invention of the first steam-propelled vehicle. The modern oil and gas industry is about 150 years old, the airline industry is over 100 years old, and the computer industry has been around 50–70 years since the invention of the first primitive computer. If one takes into consideration the number of life-saving therapeutics and diagnostics currently produced from biotechnology, then realizes biotechnology products can require up to 15 years to reach commercialization, one can see that this industry is poised for tremendous growth and impact on the future of medicine and human health. The biotechnology field is filled with amazing discoveries that continue to change how medicine is practiced. Twenty or thirty years ago, you could not find a course on medical genetics in any medical school curriculum now it is essential, as genetic discoveries daily impact the practice of medicine – all because of biotechnology.

The Business of Biotechnology Is Like No Other

The business of biotechnology is like no other. It involves all the same risks that all small businesses face, yet there are significant additional risks not shared with any other business endeavor. For all small business start-ups, sobering statistics reveal that about 80% of all small businesses fail within the first five years. In the biotech business, there is the requirement of enormous amounts of money to get a product developed. Estimates range between $25 million to $100 million dollars for medical diagnostics and devices, and up to $1 billion dollars for therapeutic products. In addition to the astronomical financial requirements, the biotechnology industry is one of the most highly regulated industries in business. In spite of these challenges, this is still an exciting time for biotechnology. Only through persistence and ingenuity, have great discoveries been transformed into products that improve the life and health of millions. Just as the serendipitous discovery of penicillin lead to the production of life-saving antibiotics – innovation, coupled to persistent wills refusing to give up, is leading to products that now impact our everyday lives – such is the future of biotechnology.

What Exactly Is Biotechnology?

The operational definition of biotechnology differs greatly depending upon whom you speak with. In the broadest terms, biotechnology encompasses any application of engineering and technology applied to the life sciences, and usually refers to the use of living organisms in the making of a product. Interestingly, the term "biotechnology" was coined almost a century ago in 1919 by Karl Ereky, a Hungarian engineer. At that time, biotechnology meant all the lines of work that produced products from raw materials with the aid of living organisms, including animal husbandry and other similar techniques.

During the 1980s, the early biotechnology industry began with the insertion of human DNA into bacteria and mammalian cell lines through genetic engineering, to produce the first human protein by cell culture fermentation. This was the bio-tool that ushered in the modern biotechnology era. Now the industry includes humanized monoclonal antibodies, genetics, cloning, stem-cells, proteomics, RNAi, miRNA, gene therapy, and molecular markers for identifying disease and predicting the future risk of disease in individuals. Biotechnology includes applications in genomics, with biochips capable of containing virtually every gene in the human body. It also includes modern vaccine development, gene-delivery, and bioinformatics. Each of these technologies is used for the development of drugs, biologics, diagnostic tests, medical devices, and laboratory services.

In addition to human health applications, biotechnology encompasses plant biochemistry and agriculture science, such as the production of genetically-engineered crops that are disease and pest resistant, even crops for delivery of vaccinations to underdeveloped countries. Biotechnology also includes the production of biofuels from various food sources. There are also industrial biotechnology applications that encompass the use of microorganisms such as bacteria or yeasts, and biological substances such as enzymes, that perform specific industrial or manufacturing processes that were not possible before. This industry is truly diverse, and comprises multiple disciplines loosely tied together under the broader umbrella of biotechnology.

Biotechnology Clusters and the Decentralization of Biotech Hubs

The size of the biotechnology industry is rapidly expanding. According to Ernst and Young[1] in 2007 worldwide there were 4,414 public and privately held biotechnology companies with 1,744 of them located in Europe. In the US, there were 1,502 biotechnology companies with 386 of them being publicly-traded – 30 years earlier there were no more than 5. These public companies alone collectively employ approximately 204,930 individuals worldwide, with 134,600 public company employees located in the US. They estimate that in 2007 there were over 300,000 employed world-wide in the biotechnology industry.

---

[1] Ernst and Young's "Beyond Borders: Global Biotechnology Report 2008".

2007 Publicly Traded Biotechnology Companies by Region in the US	Number
San Francisco Bay Area	77
New England	62
San Diego	42
New Jersey	32
New York State	28
Southeast	23
Mid-Atlantic	22
Los Angeles/Orange County	21
Pennsylvania/Delaware Valley	15
Pacific NW	14
Texas	13
Mid-West	11
North Carolina	9
Colorado	4
Utah	2
Other	11

Source: Ernst and Young's "Beyond Borders: Global Biotechnology Report 2008"

Some of the more well-known biotechnology therapeutic companies include Genentech, Amgen, Genzyme, Biogen-IDEC, and Gilead; others in the medical testing, diagnostic and medical device field include Applied Biosystems, Abbott and Affymetrix.

The San Francisco Bay Area, because of its burgeoning venture capital industry that funded Silicon Valley computer technology companies, has been a plentiful source of capital that also spawned the early development of the biotechnology industry. The New England area, with its proximity to many universities and academic medical research institutes, transformed itself into another biotech cluster that fueled the growth of more fledgling companies. As the industry began to expand in the US, we witnessed growth mostly limited within certain geographic regions of San Francisco, Boston, San Diego, Raleigh-Durham-Chapel Hill, Washington DC, and Seattle. Rapid growth of biotechnology in these cluster regions was largely due to the convergence of five essential elements present simultaneously at these locations. These five elements continue to be critical for the establishment of biotechnology clusters anywhere in the world:

1. The availability of significant amounts of venture capital willing to fund life-science companies.

2. Proximity to high-caliber academic and basic research institutions performing research in the life sciences.

3. The availability of entrepreneur leaders with seasoned experience in growing and building biotechnology companies.

4. The ready reserve of a talented scientific work force to support and grow these organizations.

5. The presence of laboratory incubator space meeting the special needs of these biotech companies.

Since these five elements were already established and growing in these hub locations, a greater proportion of biotechnology companies continued to spring up in these regions. However, as the biotech industry began to expand, and the economic value of this industry became evident, other regions joined in creating their own start-up biotech companies because good ideas are found in every state and each one has at least one quality research institution. Gradually, venture capital became less restrictive to locale, and new venture capital organizations were created in formerly nontraditional biotech locations. Many state and local governments have recognized that the biotechnology industry is one that can significantly diversify and improve the economy of their state because it has the advantage of being a clean industry with higher-than-average wage-generating potential for employees – all of whom pay taxes. As a result, we are now witnessing the decentralization of biotechnology into all regions of the US and in almost every nation in the world. Many cities and states have allocated hundreds of millions, even billions of dollars in initiatives for state support of biotechnology. The initiative fueling decentralization is the creation of the infrastructure and development of these five essential elements locally, to be competitive in building a life science industry.

For instance, in the middle of the country, the Colorado Science and Technology Park at Fitzsimons is an example of what can be done to facilitate the organic growth of biotechnology in one's own state. A decommissioned Army medical center was closed down a decade ago in Denver, and 18 million square feet is being converted to house 30,000 new employees with many of the companies being spin-off technology companies from the University of Colorado Health Science Center (UCHSC). The state's investment of $4 billion dollars will become an economic generator for Colorado and stimulate the development of many new biotechnology companies.

In an oil and gas state such as Oklahoma, the oil bust of the 1980s fueled the desire for industry diversification and helped create incentives for the development of science and technology companies. After the oil bust, the state passed an Economic Development Act that created the Oklahoma Center for the Advancement of Science and Technology (OCAST). OCAST adopted an NIH-type peer reviewed grant system that funded over $115 million dollars into science and technology research that translated into companies creating technology products developed in the state of Oklahoma. Another initiative created the Technology Commercialization Center, run by Innovation to Enterprise (i2e) and funded through OCAST. i2e provides resources to assist new technology businesses get established, evaluate business plans and even provide seed capital for the formation of these companies, and has an annual budget of over $2 million dollars. At a previous company, we received over $900,000 from OCAST's competitive grant program, and at a second company, we won another $850,000 in grants to fund our company's product development. Other nonprofit foundations such as the Presbyterian Health Foundation had a vision for the development of a new research park with about 1 million square feet of Class A wet labs and office space that provides state-of-the-art research laboratories and incubator space at significantly reduced rates to many biotechnology companies. Examples such as these are being repeated in almost every state across the US and in other non-biotechnology locales in the UK, Europe, Canada, Japan, Australia, China, and India. Over the next several decades, we will witness the complete decentralization of biotechnology into areas previously void of any life science commercialization.

Is the Biotechnology Business Model Broken?

Some critics say that the biotechnology business model is broken. They cite the inefficiency in generating significant revenues and profits from the industry as a whole, absent a handful of successes. Their argument consists of a summary of the combined revenues generated by this industry, which continues to grow, but the combined profits are at or below zero. They ignore the fact that most of the companies in this industry are early-stage development companies, which consume rather than produce capital. Evaluations like these can be problematic, because they arise from comparing the biotechnology industry to non-science-based businesses using the same metrics. If the business of biotechnology is measured across the same time-frame as say the IT or computer industry, then the critics are correct. However, the biotechnology industry has a product development time-frame that can extend 10–15 years, which significantly impacts the collective revenue and profit the industry generates during the first or second generation of companies. Lengthy product development and clinical trials coupled to prolonged regulatory approvals require many years or potentially decades to complete, expanding this time-frame.

All industries go through adaptive refinement. In the 1980s and 1990s, during the early days of biotechnology's growth, there were few examples to follow, limited experience to draw from, and no proven business models for these companies. In the years since then, more support systems are in place, more experience is available to draw from, and better tools are available, as are more sophisticated investors to help the entrepreneur achieve their goals. The biotechnology industry's ability to quickly hone in on a drug target or molecular marker with the highest likelihood of success will improve, and this will increase the success rate and speed to commercialization. The concept of improving targeting efficiency is universal and similar in the petroleum industry as they sought to locate new sources of oil. In the early days, oil companies simply dug holes in the ground in the most likely places with few successes. Today oil companies use sophisticated devices based upon sonar, and other high technology geological devices, and tools to improve the likelihood that a drill will hit oil. The big difference between the oil industry and the biotechnology industry is the length of time it takes to learn the well is "dry." Biotech companies may not discover that their well is "dry" until their drug candidate has completed phase III clinical trials, and this can be 8–12 years from start. In this industry, that amount of time can equate to hundreds of millions of dollars. Unfortunately, the steps in biotechnology product development cannot run in parallel. Just as you cannot put two 4½ month pregnant women together to get a baby – some things just require time.

If today's revenue and profits were the sole measure of an industry's value, then the airline industry is a broken business model also. However, the airline industry exists because it provides a needed service of air travel, which is not satisfactorily substituted by the automobile or railroad industry. It is true that each industry must ultimately become profitable and self-sustaining to perpetually contribute valuable products and services to society. However, the biotechnology industry contributes a value that cannot just be measured today in near-term profits. It is true that the path to biotech commercialization is littered with hundreds of companies, which we no longer know the names – yet this can be said of many other industries. However, the success of the biotech industry, no matter what the business model, will be dictated by the value of their products and services in attracting willing financial markets, that value their contributions enough to support them to maturity. As long as biotechnology products are produced with life-saving medical significance, the biotechnology industry will continue to flourish, and provide opportunity for others to develop life-saving products inconceivable decades ago. As more biotechnology companies mature, no doubt this industry will improve in overall efficiency and profitability. But more importantly, its life-saving products it will have already made a significant impact on the health and welfare of all society.

Chapter 2
What Makes a Biotech Entrepreneur?

Biotechnology products arise from successful biotech companies. These companies are built by talented individuals in possession of a scientific breakthrough that is translated into a product or service idea, which is ultimately brought into commercialization. At the heart of this effort is the biotech entrepreneur, who forms the company with a vision they believe will benefit the lives and health of countless individuals. Entrepreneurs start biotechnology companies for various reasons, but creating revolutionary products and tools that impact the lives of potentially millions of people is one of the fundamental reasons why all entrepreneurs start biotechnology companies. Certainly, biotech entrepreneurs hope to make truckloads of money by building successful companies with billions in revenue. But most biotech entrepreneurs have an altruistic streak fueling their persistence, that keeps them going through the hardships and challenges that would stop cold those just looking to make a quick buck.

As the biotech entrepreneur and the start-up team begin a journey down what may be one of the most exciting and rewarding experiences of their entire career, they live out an opportunity to impact the world by developing medical products that combat some of the most challenging and deadly diseases that plague humanity. The sentiment heard universally from experienced biotech entrepreneurs is, that starting a biotechnology company is exciting, stimulating, and frightening – all at the same time. This chapter reviews the background of the biotech entrepreneur, as well as some of the most important personal characteristics of these individuals. Whether they live in Boston or Beijing, San Francisco or Singapore, Dublin or Denver – the characteristics of biotech entrepreneurs are all the same.

A Different Breed

The biotech entrepreneur is unique from all other entrepreneurs. Do not think this a patronizing statement, rather one of the cautionary relevance. Yes, the biotech entrepreneur must still possess the same attributes all other entrepreneurs do; those being independence, confidence, a desire to take the road less traveled, a passion for their work, the ability to work long hours, unrelenting persistence, and willingness to take risks. In addition, the biotech entrepreneur is usually an accomplished scientist,

C.D. Shimasaki, *The Business of Bioscience: What Goes into Making a Biotechnology Product*, DOI 10.1007/978-1-4419-0064-7_2,
© 2009 American Association of Pharmaceutical Scientists

bioengineer, physician, or businessperson. Most often, but not always, they have a PhD, MD, MBA, or combination of these educational backgrounds. These individuals usually have well-paying and secure positions, and are already experiencing some degree of success in their current position. A biotech entrepreneur voluntarily leaves their comfortable world, and steps into an industry that carries uncertainties and risks unique to any other business.

I am convinced that only true entrepreneurs can start a life science company. Any other personality type would be too cautious, too analytical and too practical, and conclude that it is futile to even begin. This is because biotech entrepreneurs do not just see problems, they envision an endless number of solutions to any given situation (never mind that many of these require resources not available, or it may have never been done before). Do not for a moment think that the biotech entrepreneur is cavalier – they just arrive in this world with a heavy dose of eternal optimism. A biotech entrepreneur recognizes problems, but does not focus on them for long. This is a strength of the biotech entrepreneur – but it can also be their downfall if not moderated.

Know the Challenges: Count the Costs

There are many challenges that await the entrepreneur. A biotechnology company is a melding of business and science, and thus it creates a business of scientific uncertainty. The product development process contains unpredictable biological and technical risks. These risks arise from a core technology based upon promising yet unproven science. Entrepreneurs must be prepared for an extraordinarily long product development timeframe. The average time to reach commercialization for biologics, drugs, and other types of therapeutics can take upwards of 15 years to reach the market. Diagnostics, medical devices, and molecular tests can range from 3 to 7 years. There are extraordinary financial risks. To develop a biotechnology product, one must secure exorbitant amounts of capital over many years, even decades, to complete development. Depending upon the type of product to be developed, the amount of money required may range from as little as $50 million, to hundreds of millions of dollar. There are regulatory risks as well. A company cannot just produce a product and sell it as in many other high technology businesses – in the US one must first get approval from a $2 billion dollar governmental agency called the FDA, standing between their product and commercialization. During this regulatory review, clinical testing results may be scrutinized for years – most likely within a changing regulatory environment. There are also new market risks. After a product receives approval, a biotech company then faces an untested and unproven market for a product that most likely never existed before. For the biotech entrepreneur, these risks are in addition to ones all entrepreneurs in any business face, such as overstretched operational capacities, new market development issues, and challenges recruiting quality people, to name a few. By stepping into the shoes of a biotech entrepreneur, the magnitude and number of obstacles surpass those all other entrepreneurs face. *Welcome to the exciting world of the biotech entrepreneur!*

Do You Have What It Takes?

Before undertaking a start-up biotechnology company, one should count the costs because the stakes are exceedingly high. By understanding the challenges before beginning, one will be better prepared to handle them when they arise. Many individuals seek to accomplish lofty goals, but when things get tough, their desire wanes. Accomplishing anything of great value requires a significant amount of time and effort. The amount of time and effort is proportional to the significance of the endeavor.

Sometimes we want what other people have, without going through what other people went through to achieve it. A man with a broken arm was in a cast for 8 weeks. He subsequently went back to his doctor to have his cast removed, and asked "Will I be able to play the piano when this cast comes off?" His doctor said "Of course you will!" The man smiled with joy, and said "that's wonderful, because I couldn't play the piano before I broke my arm!" Be sure that you are committed to do whatever it takes to be successful *before* you start your company.

A good way to know if someone possesses the ability to accomplish things of significance is to examine their past experiences. The entrepreneur should inventory their past to see if they stuck it out when things got tough. Did they find creative solutions to solve problems? Did they demonstrate patience and persistence when faced with a difficult situation? Did they persevere to the finish?

Personal Costs

It is equally important to recognize the personal and family costs before starting down this path. The biotech entrepreneur should have a strong support system of family and close friends because their work will also impact these individuals. A support system is vital, because there will be times when their understanding, encouragement, and help (possibly even their money) is needed.

I recently met a first-time entrepreneur, who asked me to evaluate his technology and assist him with technical and market advice in applying RFID to an agriculture application. He was finishing his Ph.D., and simultaneously preparing to get married while starting his new technology business. He was very excited about the prospect of his new technology opportunity and had received some early financing interest from potential investors and industry partners. After advising him on the business aspects, I then encouraged him, if he had not already done so, to sit down with his future spouse and have a discussion about what this new endeavor will potentially require in time and financial commitment, and to be sure that she was in favor of him doing this. He said he had not thought about that, but would do so. Subsequently, months after their wedding, and many months into his business development, he came back to thank me for bringing this to his attention because it helped him tremendously. He said they did discuss these issues and considered the time and financial commitment, and then they both agreed it was the right thing to do. With that, he had her support and understanding. As the business began to consume more of his working and free time, she was committed to his efforts which enabled him to continue with confidence and support.

Four Backgrounds of Biotech Entrepreneurs

The typical biotech entrepreneur who starts a company usually comes from one of four background types. Although individuals from any background can start a biotechnology company – so long as they have the ideas, skills, and motivation. Unfortunately, the odds for success are heavily weighed against anyone not from one of these categories. Each of these backgrounds comes with their own strengths and weaknesses. The four most common backgrounds of life science entrepreneurs include the following:

1. The *Scientist/Physician/Bioengineer* who comes from an academic institution (University, Research Foundation, Non-profit Research Institute)

2. The *Scientist/Physician/Bioengineer* who comes from within the life science industry such as another biotechnology company

3. A *Businessperson,* such as a former executive in the life science, pharmaceutical or venture capital industry, who is not a Scientist/Physician/Bioengineer

4. A *Core Group of Individuals* that are spun off from another life science organization within the industry

Most biotech entrepreneurs and founders can be classified into one of these four categories. There can also be fundraising effects associated with each category of biotech entrepreneur, which we will term "Founder Effects." These are discussed in Chapter 8.

For the Scientist: Things vs. People

The largest majority of biotech entrepreneurs come from a scientific background, and this influences the things to which they gravitate. Fundamentally, a large percentage of scientists become scientists because they either enjoy or are more comfortable working with "things" (experiments) rather than "people." This does not imply that scientists do not like people; it is just that they find the challenges and rewards of research more fulfilling, compared to the challenges and rewards of predominantly working with people. Ironically, a significant part of a biotech entrepreneur's job requires that they work more with people rather than with experiments and science. As with any job there are skills that must be acquired in order to do what needs to be done whether one enjoys it or not. Many scientist entrepreneurs find they need help with their people skills because these may not have been honed to the same degree as their scientific skills. For the physician entrepreneur, this may not be an issue because they usually have clinical experience, and if they have been in clinical practice for any length of time, they may have developed reasonably good people skills. For the businessperson entrepreneur, if they have been successful at a business previously, working with people and dealing with these issues are where they usually excel.

Good scientists are capable of becoming good CEOs. An accomplished scientist and early employee of Genentech was David Goeddel. He was a significant force in the establishment of Genentech and an exceptional hands-on bench scientist. Goeddel was instrumental in successfully cloning many of Genentech's early products that are now on the market. Goeddel later went on to cofound Tularik, another biotechnology company in South San Francisco, with Robert Tjian and Steven McKnight, and eventually became CEO. In a transcribed interview, he openly admitted that he was not a people person at first, but out of the need and desire to do things well, he adapted to the requirements of a good CEO, which he found were different from the skills of a good scientist.[1] Acquiring new skills can only be accomplished if the entrepreneur first recognizes and acknowledges that they have shortcomings in a particular area. Individuals who are successful in one area sometimes have a hard time admitting they are mediocre, or down-right terrible, in another area. All good scientists have the potential to acquire other skills because they understand the learning process and the importance of training to gain proficiency. A scientist desiring to be a good CEO, can do so if they set their mind to it, avail themselves to good learning tools, and persist in learning.

For the Businessperson: Communication with Scientists

The entrepreneur with an entirely business background has a different challenge to wrestle with than does the scientist. Individuals with business backgrounds have a difficult time appreciating the technical challenges of the science, and typically have trouble understanding why technology development takes so long. This gap can be disastrous for planning. Most businesspersons do not have the background to readily understand the limitations of a technology; as a result, they may be surprised by "unplanned" scientific problems, resulting in product development delays that translate into financial shortfalls. To them, it may not seem obvious why a scientist cannot just run an experiment once, get the results, and move on.

Frequently, tensions arise between the businessperson and the scientist. This happens when the businessperson may erroneously promise things the technology just cannot perform, or the scientists may become frustrated with the businessperson

---

[1]David V. Goeddel, Ph.D., "Scientist at Genentech, CEO at Tularik," an oral history conducted in 2001 and 2002 by Sally Smith Hughes for the Regional Oral History Office, The Bancroft Library, University of California, Berkeley, 2003.

because they do not appreciate the technical difficulty, and why things take so long. Businesspersons need to develop frequent and good communication channels (not superficial) with the scientists. They should practice having shared understanding of goals and objectives, and identify the challenges for both the business and the science. It is critical that the businessperson learn as much as possible about scientific issues by asking questions of the scientists to gain appreciation of the technology and its limitations. By understanding the scientific limitations and then sharing the market needs, the businessperson may influence the scientific development in a way that better meets the market needs, while simultaneously engaging the scientists to help solve the market problems. Businesspeople may be surprised by the solutions scientists can come up with, once they understand the business and market issues.

First Time Entrepreneur, First Time in Business

Alternatives to Consider Before Taking the Leadership Role

If a professor, physician, scientist, or bioengineer from an academic background is contemplating starting a biotechnology company, but unsure about the decision to lead the company, they may want to consider a few alternatives. There are several ways to participate in the entrepreneurial process without carrying the responsibility for the entire organization and its outcome. Each of these alternatives still provides valuable experience for participating in the entrepreneurial process and equips you better for a future entrepreneurial opportunity. Options include:

1. Take the position of Chief Scientific Officer or Chief Medical Officer in the new company and allow someone else more experienced to shoulder the major business and financing responsibilities for the organization.

2. Participate as a cofounder in getting the company started and then return to research or medical practice, while contributing in a role as a Scientific Advisory Board Member or Board Member.

3. Be the scientific founder, help establish the company, and possibly work to get seed funding with the idea of later recruiting an experienced CEO. Find out how well you do before deciding whether or not to continue in this role, or turn it over to someone with more experience.

Young scientists or young businesspersons with minimal business experience may want to consider supportive roles such as these, or consider a transition role in the company rather than taking responsibility for the entire organization. In this way, they can learn by observing and participating, rather than being solely responsible for the outcome of the company. By doing this, they can gain valuable experience and apply it to the next opportunity, where they can lead with more experience and an understanding of the process and issues faced.

Although I had about 10 years of entrepreneurial industry experience when a group of four individuals cofounded a biotechnology start-up, I filled the role of VP of R&D, while the businessperson with a marketing background appropriately filled the CEO role. In my role at this company, I was still actively involved in the growth and development of the organization and participated in the fundraising and in the IPO road show process. As the company grew, I later took on the additional role of Chief Operating Officer and gained more insight into the running of a biotech business. These experiences better prepared me for the responsibilities at my subsequent biotechnology start-up, where I did assume the role of CEO. Just because the entrepreneur is not leading the organization, does not mean they will not participate in shaping its future.

An academic scientist's background and experience can sometimes be a help or a hindrance when raising money. The more practical business experience a scientist gains, the more confidence potential investors will have in that person's ability to successfully lead a company. Biotechnology investors invest in people they trust – not just in technologies alone. Unfortunately, sometimes stereotyping of scientists can occur. On one rare occasion I heard an unconventional venture capital investor angrily declare "I would never invest in a company where a scientist was the CEO". Too bad, that investor would *not* have invested early in Amgen nor invested in Genentech today because a scientist *is* the CEO. This concern should not be the sole reason for a decision whether to lead a company or not, but it is a factor that one should at least be aware of early on. Those that are good at working with people, and can communicate well the value of the organization to investors, and have a strong desire to build and lead a biotechnology company, really will not be satisfied unless they do.

Acquiring Additional Business Skills

Scientists or physicians without a business background may want to consider supplemental training in business either before, or during, their involvement with a start-up company. Even the businessperson may want to consider additional training because they are not necessarily endowed naturally with the capabilities to lead a company simply because they have a career in some aspect of business. There are a variety of options for gaining further business skills and knowledge, and these helps are offered by some of the best business schools in the country. Most all business schools offer short courses that cover such topics as Finance for Executives, Health Care Strategies, and Biotechnology Business Issues, all the way to traditional MBA and Executive MBA programs. If you are considering a major commitment to formal business school education, and you can manage the costs and time commitment, an Executive MBA program is a great way to go. This is not an endeavor for the faint of heart as it is an intense time commitment, is also costly, and can run well over $120,000 for tuition during the two years of a program. However, it provides an opportunity for a mid-career person to gain the skills and knowledge they may want without quitting your their job and taking two years off to complete a traditional program; it is also an excellent experience that will sharpen your skills as a business executive and add broad-based knowledge to your decision making. The choice of school is critical, as it will shape your learning experience and your view of business in the future. If you go this route, choose a business school that incorporates your values, and offers high-caliber instruction in a collaborative environment.

I was in the biotech industry for 19 years when making a decision go to business school. I chose an Executive MBA program to learn with peers of equivalent career experience and working business knowledge. I sought a school with a strong core curriculum and also with high ethics and leadership training, which are qualities that are needed in the business world today. The way I personally chose a business school was by attending class lectures at different institutions before making my decision to apply. The Executive MBA program turned out to be a challenging and time-intensive two years (studying an additional 20–25 hours per week on top of a full-time job, not including class time); however, it is a decision I would make all over again. Having completed an MBA, I can honestly say I make better business decisions based on the knowledge and insight gained from my professors and peers through this training. There are many excellent business schools world-wide, so if you are interested in pursuing this avenue as an investment in your future, it may be well worth the time and effort.

For the Young MBA

For the young MBA student or graduate contemplating starting a biotechnology company, they should consider acquiring some type of training in the sciences, although a formal science degree is usually not necessary. Many good business schools have biotech courses available, and some even have dual Master's degree programs in Biotechnology and Business. Whatever your scientific interest, you should avail yourself to as many good educational resources as you can find.

Be a Life-Long Learner

In reality, most entrepreneurs do not have the time to pursue a formal business education. However, they can still acquire additional business skills by developing a practice of reading good business books to expand their knowledge about business issues. During my scientific career, I constantly read books on business practices, project management, and leadership, along with biographies of successful CEOs. Another option is to take advantage of half-day seminars or short presentations to help where you may be lacking, be it finance, marketing, negotiations, strategic alliances, business development, or scientific and technical areas. One can also gain additional business knowledge by meeting with successful business leaders or other biotech entrepreneurs, who can provide practical insights about making wise business decisions. As you do this, you will find yourself thinking about business problems differently. By learning new skills and business concepts, you are acquiring the tools to help solve business problems. As the saying goes, "If you are a hammer, everything looks like a nail." Work to increase the tools in your tool box and you will be able to make better decisions. Maintain a life-long learning mindset, and it will ensure that you will not always have to learn lessons the hard way.

Essentials of a Biotech Entrepreneur

Passion and Vision

The most successful biotech entrepreneurs all possess these two components – they have a *vision* and *passion* about what they are doing. Entrepreneurial passion is not an emotional characteristic, but a driving desire to accomplish something they firmly believe in, and will do, no matter how difficult the challenge. Being a visionary is about:

1. Seeing something others don't see

2. Communicating what they see in a way that inspires others to follow

Entrepreneurs should be sure they possess these characteristics if they are going to start a biotech company, because they will need them when they face the many challenges during company development. Also, potential investors will be looking for these characteristics because they know that they are essential to building a successful company.

Assess Your Strengths and Weaknesses

The more successful an individual becomes in one particular area, the more blind spots they tend to accumulate in other areas. As a result, most people do not have a good appreciation of their real strengths and weaknesses. The best way to know one's weaknesses is to ask those who are closest, to tell you what they are; however, the problem with this is that others are rarely comfortable telling someone the truth about their weaknesses. Another way to assess your strengths and weaknesses, is to complete a personality or work behavior profile such as a Myers-Briggs, DISC, or other self-assessment test. Many times these assessments are available with an interpreter or coach to assist in the translation and discussion of the results. Completing such a test is not a requirement for starting a biotechnology company, but it will be worth the time and money to complete it sometime during your biotech career. I highly recommend that you do it sooner rather than later, because it will help you objectively learn your weaknesses. By knowing these, you can work to improve them, and later balance them by hiring staff with complementary strengths in your areas of greatest weakness.

I personally have all my senior staff take a work behavior profile test and have the results explained so that our team can understand each others communication styles, and how each naturally responds to their environment. Years ago I would have discounted this exercise as fluff, and say, "we just need to get the work done." I have since learned that a fair portion of problems arise from miscommunications, differing communication styles, and different reference points, rather than from true disagreements. Having a good understanding amongst team members results in open and honest communications. However, when team members do not trust each other, the company does not move in the most straightforward path. When team member problems occur, a typical approach is to force everyone to work together whether they like it or not. This can lead to a situation where individuals begin looking out for only themselves instead of the goals of the organization. This generates artificial problems and is counter-productive – realize there are enough real problems in starting a biotech company, without generating more. Be sure your team is equipped with good communication tools, and help them learn useful interaction skills early on. By doing this, it will minimize artificial problems later and the team will become a more productive group.

Learn New Skills Quickly

Regardless of how accomplished someone may be in their present career, they should not be fooled into believing this guarantees success as a biotech entrepreneur. Entrepreneurs need to learn new skills quickly or they will not be successful. The reality television show "The Apprentice" is a successful television series in which creator Mark Burnett brings accomplished individuals from various areas of enterprise to vie for a spot as the winner of Donald Trump's next apprentice. The latest modification to this show, featured a twist in the casting, where famous and successful individuals from various walks of life were cast for the "Celebrity Apprentice." During this particular season, 18 hugely successful individuals included: world heavyweight boxing champions, Olympic gold medal athletes, rock stars, TV stars, a super model, and successful personalities. The object was to win at the business tasks as a team: Whereas, the losing team faces Donald Trump in the Board Room, and the least successful person on that team will hear the words "you're fired!" Interestingly, two of the first three "terminated" celebrities were accomplished gold medal Olympians: Jennie Finch, the 3-time USA gold medal softball pitcher, and Nadia Comaneci, the 5-time Olympic gold medal winner in gymnastics. These two are fiercely competitive athletes, who keenly know what it takes to be successful competitors. Both athletes experienced victory because they devoted their life to self-discipline, focus, planning, and execution, to achieve Olympic Gold Medal success. However, excellence and accomplishment in one field, no matter how great, does not immediately endow one with superior business skills. There is no doubt that these athletes, with proper business training, could quickly excel and become superior business competitors, but without time and training in business, all individuals begin at remedial levels. The Biotech Entrepreneur must adopt the mindset of being a quick learner because it will serve them well as they establish and grow their business. If one quickly recognizes what they do not know – and learns what they need to know – they can avoid the pitfalls encountered by less successful individuals.

Good Business Sense

Most good business decisions are really just common sense. The problem is that common sense is not always "common." Having common sense allows one to make good business decisions *if* they understand the issues. However, if an entrepreneur has trouble understanding business issues, they will also have trouble making good business decisions. For the scientist, a good way to become exposed to business issues is to read business journals and newspapers such as the Harvard Business Review and the Wall Street Journal. If the entrepreneur finds that nothing in these publications interest them, they may not yet have an appreciation for business issues. This does not mean they cannot start or run a biotechnology company. However, someone who does not enjoy the business aspect of running a company should consider some of the alternatives to leading the business, as discussed at the beginning of this chapter.

Have the Ability to Speak Two Languages: Be a Multi-Disciplined Translator

A significant universal barrier to achieving success for a biotechnology company is the absence of a "multi-disciplined translator". A multi-disciplined translator is a person who understands the marketing, financial, and business issues, but also understands the technical and scientific issues – and speaks both languages well. One of two scenarios is usually played out in biotech start-up organizations: The first, is where the entrepreneur without a technical background, cannot or does not want to understand the scientific issues and limitations. In this situation, the entrepreneur operates in a pseudo-environment where they believe the science operates in a certain way, but in reality it does not. The second situation is where the entrepreneur is a scientist who cannot, or will not, learn to communicate scientific issues in a way that the business and financial persons and investors can understand. They tend to go headlong into a discourse on issues that are of no interest to anyone except those with a technical mind. The scientist, in this case, does not attempt to translate, but speaks to businesspeople, investors, and Board members in scientific jargon, and approaches them using the scientific method. Neither of these situations accomplishes true communication. As a result, transmission of critical information necessary for effectiveness does not occur.

All scientists are trained to be skeptical when analyzing data. Scientists are good at looking for problems, seeing issues of misinterpretation, and identifying incorrect conclusions – these are great skills for performing quality research. However, these qualities are terrible communication tools for speaking with business and financial partners. For the scientist to be an effective leader, they must have a "split brain." This does not translate to "schizophrenic." It means having a strong right brain that understands the more abstract things such as market forces, branding issues, consumer psychology, and other business development issues, *and* having a left brain that understands the science and technical issues. Whether the biotech entrepreneur has a background as a scientist or businessperson, each must speak to the other group of individuals using *their* terminology and vernacular. I do not advocate memorizing business and scientific lexicons; however, it is essential that the entrepreneur understand the concepts that drive both business and scientific decisions, and possesses a working knowledge of these terms.

When communicating, all entrepreneurs should learn to use the language of their listener. A scientist talking with a businessperson should not speak with multisyllabic words having Latin derivatives. The word "bacteria" works just fine for "bacteroidies melaninogenicus," and the single-syllable word "gene" suffices for "single nucleotide polymorphism." A businessperson talking to a scientist should not go into a lengthy discourse about "discounted cash flows" when asked "how much is the company worth?" Learn the language of the listener, and if necessary, digress to "translate" any technical concepts into words the audience will then understand. Verbal communication is simply a means of faithful transmission of thoughts and concepts to another person. Use words that best transmit these concepts to the listener. When verbal communication is an impediment, the listener never knows if the communicator's thoughts are worthy of consideration, or whether they are in disagreement, or they just do not understand. This problem is worse than communicating with someone in an unfamiliar foreign language. In the former situation, listeners comprehend the words, yet they really do not understand the context. Often they think they understand what is being said, but conclude that the communicator does not know the real issues. Practice is important to becoming a multi-disciplined translator. When communicating, ask for feedback from listeners and see if you are understood, and ask them if they can paraphrase what they heard. Most people will be patient if they know you are trying to communicate more clearly.

Successful multi-disciplined translators *think* about issues in other disciplines. Individuals who have mastered a foreign language, routinely say that they think in that language once they have become proficient. The businessperson should practice thinking about the problems scientists do. The scientist should practice thinking about the problems a businessperson faces. For those that do not know where to begin, explain the problem to a proficient businessperson or scientist and get their help. Scientists and businesspeople approach problems differently – learn to follow their thinking process, not just the end result.

Beware of the Unknown Unknowns

The biotech entrepreneur will encounter two types of "unknowns." The first is the known-unknowns. These are things easily recognized as important but you also understand your knowledge is limited in these areas. Although you would like to have knowledge in these areas, you can compensate for this by hiring a consultant or by seeking advice from those that are experts in these areas. The trickier problem for the biotech entrepreneur is the unknown-unknowns. These are things you do not know – that you do not know. The reason for this problem is you *think* you understand the issue, but you do not realize there are other hidden issues.

Unknown-unknowns are challenging to recognize because by definition one does not know that they do not know these issues. Always seek guidance and help from others, even if you think you already know the answers, because you can either learn something new or confirm your assumptions. If you practice learning from everyone, you will avoid mistakes and circumvent unknown problems. For this reason, the biotech entrepreneur needs help from experienced consultants, Advisory Board members, Board members, and experienced venture capital partners. Venture capitalists focused in the life science industry are usually familiar with most issues a biotechnology company will face, so it is wise to seek and utilize them for more than just financial reasons (more on this subject in Chapter 8). Having partnerships with larger organizations that have experience in your life science arena also assist in uncovering the unknown-unknowns. Just remember, listening to everyone is not the same as accepting advice from everyone, for truly, many individuals give advice that do not know or understand the issues you face nor do they understand the ramifications of each decision. Grow in discernment because everyone you talk with will have an opinion, but not everyone will know what is best for your situation – seek experienced opinions.

Gain Knowledge and Wisdom

Entrepreneurs are guaranteed problems and challenges, but having knowledge is not enough to resolve them. Knowledge is simply the acquisition of information. In addition to knowledge, one needs wisdom, which is the proper application of selected pieces of knowledge at the right time. A good example of wisdom applied to a difficult situation is exemplified in the Old Testament, where King Solomon was presented with the problem of two women, each claiming that a particular infant child was their own after one of the mothers' infants died in their sleep (DNA testing certainly would have helped). In this situation, one woman who was always envious of the other, accidentally rolled over and suffocated her own baby while it slept near her. The jealous woman did not want the other woman to have a baby since she lost hers, so she switched her dead one for the live one while the other woman slept. The disagreement appeared irresolvable. King Solomon wisely said that because they both claimed the baby, the only way to solve the problem was to split the baby and give half to each mother. He instructed the palace guard to draw his sword and split the remaining baby in two. The first women said "yes!" but the second women instinctively cried "give my baby to the other women for I would rather see my son live than die!" King Solomon immediately instructed the palace guard to give the baby to the second women, for she was the rightful mother.

Although this problem is not one an entrepreneur would expect to face during the start-up of a life science company, however, similar types of uncertainty can be expected. Decisions such as, which development route to take, which candidate employee will be the best hire for the organization, which venture group to go with, what is the best strategy to the market – each require a Solomon type of wisdom. Many times, additional information is found by those who will spend the time to ponder more than just the straight facts of the situation. Decisions should never be whimsical nor should they be made irrespective of the facts, but always seek as much additional information as possible before drawing conclusions and making critical decisions.

Persistence

Many times the difference between success and failure is just holding on a little longer. Thomas Edison remarked, when asked about his numerous unsuccessful attempts to find the perfect filament for the light bulb, "I never failed once, it just happened to be a 2,000-step process." For the biotech entrepreneur the saying, "success is 1% inspiration and 99% perspiration" is an important truth to remember when building a business. There will be times, when you find yourself in unfamiliar situations that are extremely challenging and have great consequence. During these times, the natural human instinct is to draw back and retreat. Retreat shows up in various ways such as shortening your work hours, avoiding decision-making, lack of confidence, apathy, shirking responsibilities, and limited communication with your team. These actions are exactly the opposite of what is needed from a leader. People do not look down on leaders who valiantly fight a battle but lose; however, they rightly fault and criticize those who shrink from their duties, when they hold positions of responsibility. During trying times, step up to face the challenge regardless of whether the answer is clear. Instead of retreating, redouble your efforts until a solution is found to overcome the situation. Investors, employees, and supporters are counting on the leader to have a plan. There is no easy route to success, and it requires commitment, dedication and just plain "hard work."

Dealing with the FDA may qualify as one of these sustained challenges, especially when you are caught in the middle of a change in FDA policy that may have been completely unexpected. This is the situation I encountered while leading a company as we sought to bring a product to the market. The first product was a multi-gene test, performed in a clinical laboratory accredited by CLIA, having regulatory requirements that were in existence for decades. The FDA changed its policy at the same time we completed our product development, which was contrary to our discussions five years previous and prior to the development of the product. In this situation, the FDA prohibited us from marketing this laboratory test without a pre-market approval, even though it was not previously required. The ensuing process, from device exemption approval, to completion of the requested studies, to the filing of a premarket application for marketing, took over 22 additional months and another $3.5 million dollars. As a result of these regulatory changes, the company was on the verge of running out of capital, and I was constantly seeking additional funding, negotiating with vendors and suppliers, and working to maintain employee morale and implement a strategy that would ultimately result in the successful mar-keting of this product. During this time, it would have been easy to give up, and some well-respected individuals even encouraged me not to waste my career and future on this battle. Ultimately we gained the go-ahead to market our first product. There may be a time to call it quits, but encountering challenges and difficulties is not a good reason to give up. By quitting too soon your life will be guided along the path of least resistance, rather than one of your choosing. It is only through overcom-ing challenges and difficulties that one can accomplish anything of real value.

Having a Deep Sense of Responsibility

Successful companies are shepherded by great leaders who understand that leader-ship is not the same as friendliness or likeability. Though most good leaders also have those characteristics, they are much more than likeable people. Deep within, they have an innate sense of responsibility for what they have been entrusted. It is from this foundation of responsibility that they make decisions, rather than from a short-sighted view of pride, personal gain, or convenience. Although individuals start and lead companies for a diverse number of reasons, not all individuals have a "deep sense of responsibility," which is essential when going through tough times if you hope to come out successful at other end. Without a deep sense of responsibility, individuals will run from difficulty or choose to ignore problems rather than deal with them when they are easier to handle. Without this sense of responsibility, uncertainty will surface in a team when the company encounters rough times because they do not have confidence that their leader will deal effectively with difficult situations.

Do not be fooled into believing that starting and building a business is all fun. Some of the most challenging times lay ahead of those who choose this business endeavor. However, some of the most rewarding and fulfilling times also await the ones that have a deep sense of responsibility, and persist through these situations rather than run from them. Entrepreneurs need to assess whether they have a deep sense of responsibility by examining how they have dealt with difficult situations in their past. If they find that they did not have this responsibility factor, all is not lost. Consciously be aware of this, and when encountering troubled times, work at doing the opposite of your natural inclination. When you make an effort to change and do something different, you will see different results. A good definition of craziness is, "doing the same things over and over but expecting different results." A deep sense of responsibility is one key component that supports other qualities commonly associated with outstanding leadership such as dedication, personal integrity, persistence, and vision.

Be a Negotiator

This may sound like an unusual topic for a biotech entrepreneur. Is not negotiation what one does when there are differences between groups of people and between organizations? Yes – but formal negotiation is only a very minor facet of negotiation. Whether we realize it or not, we negotiate many times each day, even outside of the work setting. Let us say we have plans for Saturday morning and our spouse says on Friday night, that both need to go shopping on Saturday morning for a Birthday present for a family member. One might instinctively say, "how about if I finish up what I had planned by 2:00 PM and we meet at the mall at 2:30 PM?" This is a negotiation. Let us say we rush to drop our clothes at the cleaners because we need a suit or dress cleaned for a dinner meeting tomorrow night, but the sign says "deadline for next day service at 8:00 AM". You walk in at 8:15 AM and say (very nicely) something like "I know you have a 8:00 AM deadline, but I really need these by tomorrow but got tied up this morning, is there any way you could possibly get these cleaned by tomorrow? Some type of pleading dialogue may occur and you may arrive at an acceptable solution that accomplishes your needs, which allows them to also be supportive of the outcome even though it may not be what is routine. These are examples of everyday negotiations. In your start-up company, you will find yourself negotiating each day for deadlines, deliverables for projects, balancing workforce assignments, and personnel issues. To be a good negotiator, you must understand what are the essential and the nonessential issues and objectives, and learn to concede the nonessential ones.

Enlarge Your Circle of Influence

It is important for a biotech entrepreneur to build a circle of influence by including many people they trust – these will be individuals from whom they can later seek help, advice, and assistance. When encountering new problems, it is sometimes challenging to find a solution because we often do not have the necessary frame of reference to relate. The best suggestion is to find advisors who have gone through the building of life science companies successfully, and seek their advice. Get input from individuals who are trustworthy. Do not blindly operate on the advice of someone who has never been in a similar situation, nor dealt with a particular challenge in question, but seek the advice of many counselors.

On Creativity and Change

One of the strengths of a start-up biotechnology company (yes, they have strengths) is their ability to respond quickly to change. The biotech entrepreneur's ability to respond to change is critical to the future success of the organization. Always strategically assess the environment, the market, and the opportunity for changes. Large corporations cannot make changes quickly; they have too much infrastructure and too many reporting relationships to make changes happen quickly. Learn to monitor the changes in the business environment, and then appropriately adjust the business strategy or development plans or even the market strategy to take advantage of any new opportunities.

A former medical director of a previous start-up biotechnology company I cofounded, had left to join a very large company that was quite successful. About a year later, he called back to catch up on our progress. I heard him lament that he missed the daily meetings we had where we would have a great idea, and after discussing things, go out and adjust the business strategy that same day. He said, now they have major committee meeting and subcommittee meeting to discuss things but the inertia is so great that new ideas get pushed into more meetings and more subcommittees until it becomes very watered down, if it gets ever gets done, it is months later and not what it started out to be. Although it is never a good idea to have a new "five-year" plan each week, do not lose the ability to be creative and execute quickly on your developing ideas; this is a strength of a small company.

Leadership and Core Values

The foundation of a good leader is their core values. Any individual can be a leader, but having the right core values will lead an organization down the right path. Hitler, Lenin, and Marx were arguably examples of effective "leaders," since leadership is often defined as the ability to marshal resources, command respect, share a vision, execute a plan, and inspire others to follow. These individuals are negative example of good leadership because they were based on misguided core values. Good leadership must be placed in the context of proper core values to build a successful, sustaining organization. The leader's core values are the compass by which the organization will be directed. Be sure to examine core values, because these are the principles and beliefs one reaches into when they do not know what to do, and are faced with difficult decisions. There are many great books on leadership that would be helpful to the biotech entrepreneur. For those who do not know where to begin, I would recommend Jim Collins' book "Good to Great"[2] as a place to start.

---

[2] Jim Collins, 2001. "Good to Great: Why Some Companies Make the Leap...and Others Don't," Collins Business, New York, 300 pp.

Summary

Successful entrepreneurs possess a diverse list of characteristics that add significance at various stages of their biotech career. They are multi-taskers, or as one serial entrepreneur put it – they must have "rapidly adaptive, serial, single-pointed, focused attention." First time entrepreneurs who have started and built companies will point out that the qualities described herein are ones they all wish they had a good dose of early in their tenure. A biotech entrepreneur may not initially possess every skill required to lead a talented and diverse team to build a successful business, however, as with any skill set, most can be learned if the entrepreneur is willing. Once these skill sets are acquired and proficiency is gained, these entrepreneurs generally go on to start many other biotech companies. These serial entrepreneurs successively build upon each previous learning experience and become proficient at their work and at recognizing the critical components of a successful business opportunity.

Whether someone seeks to start or join a company at any stage of development, a good analogy for building a biotech company is to realize that every great architectural structure we see today was first conceived in someone's mind. The Eiffel Tower was just a concept – a grand idea that existed many years before the first steel girder was ever set in place. During its inception, the visionaries encountered seemingly insurmountable challenges, and were faced with many reasons to quit. Some of the naysayers of this project voiced that it costs too much, the design cannot support its weight, and the land where it is to be erected is not solid. Through the perseverance of the Tower's visionaries and their ability to communicate the vision effectively to others – they recruited a team that later solved the engineering, construction, financing, and political challenges. As a result, this monumental piece of architecture stands today. What is your vision? Communicate it frequently to others, and before you know it you will see a company established and growing through the efforts of diverse individuals, all joined together to accomplish something far greater than anyone can accomplish alone.

Chapter 3
Start with the Idea: Licensing and Protecting Core Assets

Where do biotech product ideas come from? How do biotech companies get started? Biotechnology companies typically spawn from basic research ideas originating at universities or nonprofit research institutions. Product ideas arise when scientists and professors conducting basic research envision that some aspect of their scientific work could possibly lead to a potential biotechnology product. Research is perpetual; therefore, universities and academic institutions usually have multiple licensing opportunities looking for a good home. Occasionally biotech companies also arise from spin-outs and technology licenses with origins from other biotechnology or pharmaceutical companies. Wherever these ideas originate, all biotechnology companies are usually conceived on the foundation of a good product idea. To move forward in the company building process, a license must be secured from the organization that owns the intellectual property.

What to Consider Before Licensing

Is There a Real Market for This Product?

The question that needs to be answered before considering the licensing of any technology – is there a viable market for this future product? Even though the technology may be world-class, and the potential product seems "useful," this does not guarantee that anyone will want to purchase the product once it is developed. For many biotech entrepreneurs, the nuance of market interest does not seem very important at this early stage. This is a grave mistake to make because the degree of market interest impacts a host of other issues – especially the ability to raise capital. For example, suppose there is a licensing opportunity for a novel and effective peptide technology that possesses powerful antibiotic properties, which works against a limited number of gram-positive bacteria. The science may be novel and the results may be fantastic, yet there are dozens of antibiotics on the market that are very effective against gram-negative bacteria and certainly economical to use. This project's lackluster market appeal make it difficult, if not impossible,

C.D. Shimasaki, *The Business of Bioscience: What Goes into Making a Biotechnology Product*, DOI 10.1007/978-1-4419-0064-7_3,
© 2009 American Association of Pharmaceutical Scientists

to generate much interest from biotech investors which are needed to develop any product through to commercialization. However, suppose the same technology produced a peptide with antibiotic properties against methicillin-resistant *Staphylococcus aureus* (MRSA). The market for such a product would be tremendous because of the lack of effective antibiotics that work against MRSA. Raising money for this second product would be less difficult than the first. Always realize that no matter how stellar, novel, exciting, and ground-breaking the science, if there is not a significant and viable industry market for the resulting product, the endeavor will be futile. This is also true for enabling technologies such as drug delivery or drug discovery tools – there must be a viable industry market having a sustainable unmet need, with a desire to purchase these products or services.

Before licensing any technology, it is important to first know three things about the market potential for this future product:

1. *How big is the estimated market for this product?* Is the market potential at least a billion dollars? Large markets help attract the types of biotech investors necessary to fund development of these products. If the technology is for a medical device, or molecular diagnostic, does the business have clear acquisition potential from corporations in this industry? If your product market is for a niche – can it obtain orphan drug status? Know the size of the potential market and make sure it is attractive enough for the type of investor interest needed before licensing any technology.

2. *Is there a significant unmet market need for this product?* Even more important than the market size, is having an acute need for such a product. If customers are satisfied with a current product in a category, there will be no demand to purchase a new one (a discussion of biotechnology product customers is found in Chapter 7). Always examine the potential indications-for-use of a product idea, and be sure there are not already satisfying medical alternatives, or at least be sure the alternatives are inadequate or minimally effective in satisfying the needs of potential customers. Also, it is extremely difficult to have a successful product that first requires creating a consumer demand, as opposed to just meeting an existing unmet medical need. Creating demand requires educating the customer that they "need" a product when they do not think that they do. This task is challenging and requires much time even for established pharmaceutical companies with endless resources.

3. *Is the field highly competitive with minimal differentiation between products?* Be sure that the market for this product is not currently saturated with good competitive products. If there are several strong competitors with products in this field, each with minimally different benefits, this is not a desirable market to enter. Companies with the most potential are working in fields where very few, if any, existing competitors are, or they are targeting markets where the competition does not have products that fully meet the market needs.

Even though a product idea may fit the above three criteria, there still can be a potential problem with being too far ahead of any market need. Successful product ideas do not run too far ahead of a market. Technology usage, customer interest, medical practice, and societal norms are constantly changing, and it is difficult to predict where medical needs will be, far into the future. Prematurely needed products have slow medical adoption. These types of products may also generate unresolved ethical concerns or they may face unknown future consequences that limit their acceptance. Gene therapy is one example of a technology that has tremendous potential and will ultimately be effective for many debilitating conditions – but today it is still too far ahead of medical adoption. The timing for a new product is just as important as the product application itself. Evaluating the market need prior to licensing a technology will greatly improve its chances for success. In Chapter 7, we review in more detail the market strategy, its development and impact on raising capital.

The Quality of the Science

Having a sustainable and viable market need for a product is critical, but simply identifying an opportune market will not alone ensure success. The quality of the scientific work, its technological merit, and feasibility in producing a product is also critical. The initial value of any company (and scientific team – discussed in more detail in Chapter 11) will be based on its scientific credentials and the ability to produce a superior product for an unmet market need. Therefore, if the science is mediocre, the company will have minimal value. If the science is top-tier, then the technology can be a great asset.

In assessing any technology, it is important to conduct a thorough review of the science before proceeding with a license. The best way to assess the quality of the science is by reviewing published research in peer-reviewed journals describing the technology or product concept. Licensing institutions should direct you to sufficient publications in reputable journals that support the science. When conducting a scientific review, ask to speak to the scientist(s) who performed the research because there is valuable insight gained by simply talking with them. If the technology is owned by a biotechnology or pharmaceutical company, realize they have less motivation to publish their research compared with academic institutions. Still, peer-reviewed publications remain the standard for assessing the credibility of the research and technology leading to any product.

Another way to assess the scientific merit and quality of the research is to review any grants awarded to the researchers for their work. Having multiple competitive federal agency grants supporting the research is a good sign that both the researcher and the research have merit. Competitive peer-reviewed grants are difficult to obtain and the granting process only awards funding to the top decile of proposals. In addition, a good scientific review needs to include discussions with academic and medical opinion leaders who work in the field of the license of interest. Many biotech business ideas may appear intriguing on the surface, but digging deeper into the technology may uncover other challenges that must be solved before the product could be feasible. For instance, peptide or protein-based drugs that must have oral administration have delivery issues because they are degraded in the intestine, also drugs that may be effective in the brain may have issues if they cannot cross the blood–brain barrier. A detailed analysis of the scientific merit of a technology is essential during a scientific review. Spend adequate time evaluating the science and consulting industry and clinical experts in the field.

An entrepreneur who is also the inventor of a technology still should seek outside assistance in getting an unbiased opinion about the technical merits of their research. It is very difficult, if not impossible, for a researcher to critically evaluate their own research in such a manner. Professors and scientists, in this instance, can ask the technology licensing office at their own institution, who to contact or seek help from with this type of evaluation and review. Often, university licensing offices conduct this type of evaluation while preparing to solicit outside interest. If the institution does not have this type of support, locate a businessperson, or a consultant, possibly a venture capitalist that specializes in biotech products, who can help assess the merit of the research in leading to a viable product.

Is It a Core Technology or a Single Product?

Once a conclusion has been drawn that the science and technology are of excellent quality and merit, examine if it is a platform technology or a single product idea. Why is this important? The ability to attract capital is impacted by the likelihood that a technology can produce multiple products from the same research and development work. Generating multiple products from the same technology increases interest from investors because they know follow-on products will grow the business in the future. Investors understand that a one-trick-pony company has limited chances of long-term success.

Apparent single product ideas may contain a technology platform within them, which may not have been obvious initially. It is possible that the product development process may be a technology platform that can generate complementary products. For instance, novel ways of producing humanized monoclonal antibodies could be a core technology applicable to many diseases for the development of multiple products, but the focus may have been on the prototype antibody with unique properties for the treatment of one disease. In another example, the ability to rapidly produce vaccines in mammalian cells in a fraction of the normal time is an enabling technology, but the focus may have been one application for the production of a hepatitis vaccine. In the molecular testing industry, the ability to use a combination of genes or gene-signatures to assess the recurrence of breast cancer is a single product. However, the know-how generated during the development of the first product can be used to speed the development of follow-on products for colon and prostate cancer recurrence. Do not overlook the know-how generated during the development of the first product as value-added in accelerating subsequent product development opportunities.

If the licensing idea is truly a single product with one application, then it is possible that the exit strategy for the company may be to license this product at a particular development stage to a larger company with complementary products in that market. This does not mean that having a single product company is bad, or that the idea will not get funded. In fact, if the product is successful, many times the likelihood of finding other applications for that product may improve. By spending time talking to others in the field such as scientists, physicians, clinicians, medical specialists, hospital staff, and laboratory directors, these discussions will help identify more useful applications of a technology. The only caution here is to be careful about disclosing any confidential information that has not yet been protected by patents so as not to risk inadvertent disclosure of the intellectual property.

Great Technology is Associated with Great Scientists

If the science is of high quality, it is always associated with high quality scientists and technical staff who intimately know the technology. In reality, the science and the scientists are not mutually exclusive. Most of the time, especially early in product development, the technology remains viable so long as those who have developed it remain associated with its development in some capacity. There are many ways to work out arrangements with the discovering scientists even if they do not plan on being cofounders or employees of the company. Some options have been discussed in Chapter 2 and others are discussed in Chapter 11. A technology can come with a team of *essential* personnel, *skilled* personnel, and *commodity* personnel. Essential personnel can be defined as those that, if they leave, they take some nonreplaceable skills and information with them. Skilled personnel are individuals who are hard to find but can be replaced if necessary, though you may not want to do this. Commodity personnel are individuals who are good, however one can find others and train them to do the same things in a reasonable period of time. At a minimum, it is imperative for a period of time to retain essential personnel with any technology license.

What To Do Next?

Once the final decision is made to move forward with a license to a technology or product idea, certain rights must be obtained from the organization that owns the intellectual property (IP). Most academic institutions have a Technology Transfer Office or Technology Licensing Office with staff, who are well-versed in moving science and product concepts to invention disclosures and shepherding them through the patenting process. However, there are significant differences in the sophistication and efficiency of the licensing process from one academic institution to another. One can find out what to expect from a particular institution by talking to other entrepreneurs who have licensed technology from the same institution previously. By contacting the Technology Transfer Office and asking questions, one can discover the helpfulness and expertise of their staff and quickly learn more about their licensing philosophy.

What Is Intellectual Property?

IP is the cumulative know-how to make or develop a product and its uses, which include patents, trade secrets and trademarks. IP may also include copyrights, but these are not central for biotech product protection. The IP portfolio represents the boundaries of the company's rights in a certain field. IP is a key to protecting a future product market; it is also a key consideration for investors when entrepreneurs are seeking capital to support the development of this technology. Normally, licensing is for issued patents, but in some cases licensing will involve "pending" or filled patents. Contractual licensing terms for pending patents will be different compared to licensing terms for issued patents. However, until a patent is issued, it is impossible to be sure that there is a license to anything (investors are also aware of this fact). Many times licensing fees are deferred for pending patents until the patent issues, but these are things to discuss with a patent attorney and with the licensing institution. Many individuals assume that patents are the sole and most protective barrier in slowing down competition. Having novel IP protection is only *one* aspect of a company's competitive strength, and should never be the only one. However, it serves as the foundation upon which to add other protections that support a product's market advantage.

What Is a Patent?

A patent is, in a sense, a contract between the inventor and the government. In exchange for a period of exclusivity, the inventor discloses the best method to make, use or practice the "art" they have conceived. Patents fuel the continued innovation of new ideas, and patents are incentives for individuals to create new and novel technologies that have utility. In countries where intellectual property laws are not strong or the patent laws are not enforced, one typically sees a lack of innovation and technology development.

Legally, a US patent is a property right given to the holder "to exclude others from making, using, offering for sale, or selling" their invention in the United States, or "importing" the invention into the United States. However, a patent does not grant the holder the right to make, use, offer for sale, sell, or import anything – but grants the right to *exclude* others from doing so. This may sound backwards because it means an inventor may have the right to exclude others from practicing their patent, whereas they may not have the right to practice their own patent. The reason for this is that many times patents overlap in technology. When this happens, the inventor or licensee must obtain a license from other patent holders to be free and clear to practice their own patent. Using a simplistic example, let us say your friend has a patent on the invention of the knife blade – but you hold a patent for the invention of the pocket knife. Your friend can practice his invention for making, using, selling, and importing the knife blade, but you could not practice your invention for the pocket knife without a license from your friend, because his knife patent prevents you from making, using, or selling your pocket knife.

Once a patent is issued, it is the responsibility of the patentee to enforce the patent. The life of an issued utility patent is now 20 years from the date of filing with the USPTO (plant and design patents have different durations). Prior to June 8, 1995, before the new harmonization laws were effective, a US patent's life was 17 years from the date the patent was issued. The effect of this change can be dramatic for some products in the biotech industry, because many patents take more than 3 years to finally issue and patent process delays can cut into the remaining life of the patent. Moreover, the real situation for drug development companies is the lengthy time for drug development and the completion of clinical testing and the FDA approval process. Once the product reaches the market, the remaining patent life could potentially be reduced to 5–10 years.

How Is a Patent Obtained?

It is helpful to have a good understanding of the patenting process, because this is an ongoing endeavor and the start-up company will be applying for more patents over the course of their product's development. When applying for patents, it is important to know what aspects to protect and which ones to ignore. Determining which components to protect is the role an experienced patent attorney will play as the company's patent portfolio strategy is being developed. The purpose of this overview is not to teach patent law or claim expertise, but merely to provide basic and general background into IP protection. For specific matters, it is important to consult a good patent attorney. Some of the information contained herein is also readily available on the USPTO website.

In order for any patent to be issued, the invention must meet the criteria of novelty, nonobviousness, and utility. Also, for a patent to be issued it must also be adequately described and have specific claims.

- **Novelty:** This means that the invention or idea is new and not disclosed in the public and there is no "prior art." This also means that the inventors themselves have not prematurely disclosed the idea prior to a timely filing of a patent on the invention. The essence of this requirement is that patents are only granted to ideas that are truly new. It goes hand-in-hand with the bargain theory, as there is no value to the public in granting exclusive patent rights to an invention that is already known.

- **Nonobviousness:** This just means that the idea or invention cannot be obvious. Even though the idea or invention may be novel, if it involves the combining of ideas or slight modifications of things already known, then the invention is obvious. The requirement for obviousness is specifically applied to a person having ordinary skill in the area of technology related to the invention; otherwise all ideas at some level of intelligence and creativity would be obvious.

- **Utility:** This just means that the invention must have a useful purpose. Normally this criterion is not usually a significant issue for the biotechnology inventor.

- **Adequately described or enabled:** This requirement is met when the invention is detailed in a way that one of ordinary skill in the art can then make and use the invention.

- **Claims:** These are specific statements that get to the core of the invention, and define what the inventor is claiming ownership of, in clear and definite terms.

The actual filing of the patent in the US is not expensive; the filing fees for small entities (individuals or companies with fewer then 500 employees) are currently $100 for a provisional patent, and $150 for a regular filing; a search fee costs $250, and an examination fee costs $100. The patent issuance fee is $650. The most significant patenting costs come from an attorney's time to prepare the patent application and later prosecute the patent. On average, the costs to file a biotech patent, including attorney time, patent search, examination fee, and filing fee, can run from $5,000 to $12,000 depending on the difficulty and the per hour rate of the patent attorney. For highly complex patents, attorney fees may run up upwards to $40,000.

Types of Patents: Traditional, Regular (Nonprovisional)

The traditional or nonprovisional patent is the typical patent application route. This begins with the submission of all the required information supporting novelty, nonobviousness, and utility, along with proper enablement and claims to the invention. Once a patent application is submitted, it is assigned a serial number and transferred to a patent examiner, knowledgeable in that particular field. The patent office will perform a complete search to uncover any related patents or prior art. The patent examiner will then review the information and provide an opinion about each issue that was uncovered, and its relevance to the patent application. The First Office Action, which is the first written formal response from the patent office after the filing of a patent, typically results in a nonfinal rejection of the patent, with reasons supporting the rejection. Usually, the time from filing to receiving the First Office Action for biotechnology patents is about a year or longer. Rarely will a patent be issued upon First Office Action. At this point, the inventors and the patent attorney must formulate responses supporting why the invention is novel and nonobvious. This response time period allotted, is three months, or the applicant can pay an additional fee to get an extension of three additional months. It is not uncommon to go through several addition rounds of office actions. This exchange process between the patent office is called "patent prosecution" and should not to be confused with "patent litigation," which is the process of asserting or defending an issued patent against someone believed to be infringing the company's patent. Usually this exchange results in the narrowing of claims from those that were submitted in the original patent application.

At the conclusion of the patent prosecution phase, the USPTO will either send a Final Notice of Rejection, or a Notice of Allowance. A Notice of Allowance means the patent will be issued after paying the issuance fee, and will be granted within three months. Even with a Final Notice of Rejection, there are several routes to go that may still end up with an issued patent such as a continuation, continuance, or even an appeal process. The length of time patent prosecution can take until the patent issues will vary greatly depending on the type of patent and the review backlog at the patent office. In general, biotechnology patents can take from three to five years from filing to issuance. According to the General Accounting Office (GAO), the success rate for obtaining an issued biotechnology patents is approximately 50%. These odds of patent issuance are typically supported by having a skilled and experienced IP attorney. Since most patent filings will be on inventions that contain complex biological processes, finding a good patent attorney with duel expertise, such as a degree in a pertinent scientific area and patent law, is essential. These dual degree professionals help with advice and strategy on adding to the patent and its claims, rather than just submitting the information given to them.

Types of Patents: Provisional

Provisional patents came into existence in 1995 when the USPTO offered inventors the option of filing a lower-cost first patent filing option. Once a provisional patent is filled it remains pending for 12 months and must be followed by a regular (nonprovisional) patent filled before the 12 month expiration in order to gain the benefit of the earlier filing date. The Provisional Patent provides one additional year to the potential patent coverage of your invention. The 20-year patent expiration date would begin at the time of the filing of the regular patent but have the benefit of the earlier provisional filing date. The provisional patent is also good to use when companies anticipate having additional data to support the traditional patent within a 12-month period. Other advantages of provisional patents include the ease of filing, secrecy and lowered cost. After filing a provisional patent or regular patent, the applicant can then attach the "patent pending" claim on their invention.

Types of Patents: International

In addition to patenting in the home country of origin, applications should be filled internationally to protect a product market outside of its home country because most all biotech products have international markets. The licensing institution of the technology may have already filled international patent applications depending on how long the technology has been in development. Filing under the Patent Cooperation Treaty (PCT) is a convenient way to initially get most industrialized countries covered. Companies in the US typically file a PCT one year after filing for US patent protection. The PCT will generate a patent search and written opinion on patentability for all the PCT countries. The costs of a PCT filing including attorney's fees can run from $4,000 to $8,000 or more. Once a PCT is filled, 30 months is allotted before a decision to file a patent in individual countries covered under the PCT is required. In some cases, there are fee extensions which will allow one to postpone this decision for a period of time. Once individual country filings have begun, – fees add up astronomically! Contrary to popular belief, there is no one "international patent." Translation fees, foreign attorney's fees, local attorney's fees and the actual filing fees easily can reach $30,000 to $50,000 or more. Be sure that patent filing of the IP in each of these other countries is essential before proceeding, as multiple patents at multiple stages reaching international filings can add up to significant financial requirements, not to mention ongoing annual maintenance fees.

As an example, at one early stage biotech company the accumulated patent prosecution and filing costs for the IP we licensed exceeded $400,000. When seeking a license it is possible to work out arrangements with the institution that holds the IP to allow payments on past expenses or only pay them upon reaching significant funding events. By doing this, up front fees can be reduced to lower the financial burden of past IP expenses, as there will also be ongoing patent expenses.

Get a Freedom-to-Operate Evaluation

A freedom-to-operate (FTO) opinion is an assessment by an experienced IP attorney who completes a patent search on the technology/product area and determines whether any other patent licenses may be needed to freely operate in the area of this invention. This type of evaluation is essential to have before finalizing any IP license from an academic institution. FTO assessments can be brief or very comprehensive – which translates to varying costs. Consider conducting some level of FTO that is financially feasible in order to get a good assessment of the IP landscape and this technology's place within it. The costs for a FTO opinion can range from $5,000 to $50,000 or more. During a FTO assessment, it is not unusual to find out that other patents from other institutions need to be in-licensed in order to have Freedom-to-Operate. Knowing this up front will help build the strongest possible IP portfolio, which protects the technology later during commercialization, and is essential when raising investment capital.

Trade Secrets

A trade secret is IP that is only known by certain individuals within a company but not known by the public or by organizations that may be competitive. To be considered a trade secret, most laws require that this information be of the type that derives economic value from the fact that it is secret (e.g., secret formula, customer list, best practices, etc.) A Trade Secret is another way of protecting company IP rather than filing patents. The exact recipe for Coca-Cola is a trade secret known only by a handful of individuals within the organization. Patenting such a recipe would not be advantageous to the Coca-Cola company, because it would require a complete disclosure, which would expire 20 years after it was filled. The value of a trade secret is that it could conceivably be kept in perpetuity.

In the eyes of investors, the value of a trade secret in the biotech industry is different from the value of an issued patent. This is because issued patents can be held by the organization and licensed, whereas trade secrets walk out the door at the end of every day. As a result, most biotech IP is composed of patents. However, certain aspects may be best held as trade secrets if there is no real advantage to pursuing a patent. Information that may be worth keeping as trade secrets may be formulas or processes that are not unique enough, but may be significant in giving the company an edge in their product or market. However, patents on new or unique formulations for existing FDA approved drugs are useful in extending the patent life of drugs. Other considerations for trade secrets are specialized techniques, and usages of software and other proprietary, but not easily patented information. The key is to understand the advantages and disadvantages of a trade secret on the information you want to hold exclusive to your organization. There is no legal filing for a trade secret; it just needs to be made known that this information is being held as a trade secret within the company. Discuss these issues with a patent attorney.

Trademarks

Trademarks are valuable assets for companies that have worked diligently to develop a successful brand. Trademarks are things customers use to associate a particular value or benefit to a company or its products. The USPTO defines a trademark as "a word, phrase, symbol or design, or a combination of words, phrases, symbols or designs, that identifies and distinguishes the source of the goods of one party from those of others." Trademarks must be shown to be in use to maintain them. Well known everyday examples include the Disney characters and the partially-eaten apple logo for Apple Computer. Most company logos are also trademarked or service marked. A "service mark" is the same as a trademark, but refers to a service or intangible activities rather than natural goods or products. A trademark can be protected indefinitely as long as it is issued in commerce and the renewal fees are paid.

Technology Licensing and Negotiations

So now that we have determined there is an unmet market need for the technology, and we know that the science is world-class, and we have a FTO opinion – now we need to negotiate a license to the technology. A good way to prepare for a licensing negotiation is to find out what other licensing deals looked like in the technology area that is being considered. Finding this information can require a few key contacts and some creativity because this type of information is not always common knowledge. Knowing someone with access to venture capital deal databases is helpful because some of this information may be found there. Armed with conventional licensing terms in a particular field, and knowing what the past deals looked like from the particular institution holding the license, greatly helps in securing a fair licensing agreement.

A license negotiation should not be likened to the purchasing of a used car, where the buyer may never see or deal with that group again. Seek to establish a good working relationship, with the licensing institution and an enduring one, because it is likely their help may be needed in the future. It would not be unusual to have to call upon them to renegotiate some aspect of the license later in order to accept terms of financing, or to form a future partnership that may present itself down the road. When negotiating, the goal is to establish a good working relationship, and simultaneously find the terms that allow each party to feel good about the deal they consummate. In addition to securing a license to the protected IP, having a commitment to provide any know-how that they have is also beneficial. The better the working relationship, the more diligently an institution will work to provide any and all information during the subsequent development of this technology.

Most technology license offices really want to be sure that the license goes to the licensee having the best chance to successfully commercialize the technology because their largest revenue source (royalties) will not be realized if the technology does not get commercialized. Since large amounts of capital are required for a biotechnology product to reach commercialization, sometimes an institution may want to first shop the technology to larger established organizations to see if they have interest. If such company interest is not found, they may then have a practice to license it to the inventor if he/she is interested. Although universities are being called upon to be engines of economic growth, they may have incentives to license the technology first to a local start-up that will be creating local jobs. Throughout this chapter, we predominately have referred to licensing technology from an academic institution rather than licensing from another company, although similar steps in the due diligence process should be followed. Negotiating a license from a university or academic institution may come with perceptions of a long, arduous, and drawn out process. In some cases, this may be true because not all universities have the staffing and personnel well-versed in negotiating various types of licensing agreements. However, more universities have strengthened and streamlined their licensing process as they understand the value a licensed technology can bring to their institution. Being prepared and having background information as to conventional deals negotiated for similar types of products will go a long way to accelerating the negotiating process.

What Does a Typical License Structure Look Like?

The financial terms of a license agreement can include one or more of the following from the licensee:

- Upfront payment(s): can be any amount, but usually includes all the costs incurred by the institution while patenting the invention
- Payments based upon progress milestones: these are value-enhancing stages the licensee reaches during product development

- Royalties and/or minimum royalties generated on sales of the product
- Licensing maintenance fees
- Sublicense fees
- Assumption of ongoing patent expenses

Other nonfinancial requirements the licensor may require:

- Rights to any refinements or new IP based upon the licensed technology
- Best efforts to commercialize the invention
- A limitation or restriction on the ability to sublicense the invention to a third party
- A limitation or restriction on the ability to assign the license to a third-party

For a therapeutic or biologic product, the royalty terms can range from as low as 3–12% or greater. For diagnostic or medical device products, royalties can range from as low as 0.5–7% or greater. If the IP is not core, but viewed as supportive, it will command lower royalties than for a product that is totally dependent on the license.

As a start-up company, the ability to have some flexibility in deferring fees and modifying typical licensing schedules based upon your needs and situation will be beneficial. Since the institution wants to be sure that the entrepreneur has a good chance of success, it is important to come with well-thought out plans for development prior to beginning any license negotiations. Licensing institutions know biotech entrepreneurs have lots of determination, but not much money. By presenting a well-thought out plan for product development and dealing realistically with the institution's issues of concern, many times they will be flexible to the company needs.

Factors in Licensing Negotiations

The financial terms of a license vary greatly depending upon whether the invention is a therapeutic, biologic, diagnostic, or medical device. For therapeutic and biologic products, the royalties are generally higher than for a diagnostics or medical devices.

General ranges for some license terms

	Therapeutic/Biologic	Diagnostic/Medical Device
Typical up front license fee	$25,000–$1,000,000	$25,000–$50,000
Potential milestone payments	Varies from $25,000 to multiple millions based upon the stages and technology	Varies from none to millions depending upon the technology
Estimated royalties	5–12%	0.5–7%
Estimated minimum annual royalties	$50,000–$1,000,000 or more depending upon the technology	$20,000–$100,000 or more depending upon the technology

Royalty rates can vary depending upon:

1. *The development stage of the invention at the time of the license:* Technologies that have been developed to a later product stage deserve higher royalty rates than earlier stage inception opportunities because of greater progress achieved and lowered risks for commercialization

2. *The type of license required:* Exclusive or Nonexclusive

3. *The competition in the field for the product this technology will produce:* The greater the competition in the field, generally the lower the value of the license, unless it is vastly superior to any other product marketed in that field

4. *The importance of the IP to the company producing the product:* If the technology is noncore because it may be a peripheral patent needed for FTO, there may be less value placed on the IP than if it is a core technology upon which the product is dependent

Exclusive Licenses

A desirable method for a biotech company to obtain IP is through direct assignment of the patent, but most universities and academic institutions rarely assign IP to any company. Assignment is an unconditional transfer of title along with all subsequent rights to the IP. There are several reasons why an institution will license rather than assign, including (1) Federal restrictions for the university on assignment, (2) a hedge against risk of failure of the company, and (3) a desire to continue to research in the relevant areas. The conventional method in which biotech companies obtain IP rights is through licensing agreements. These license agreements have conditions and requirements that the company must meet to maintain the license to use the IP. There are also penalties or revocations of the license if certain conditions are not met during the course of the license.

Licenses can be exclusive or nonexclusive. An exclusive license is just that – exclusive. This means that there is only one license given and only the licensee is entitled to use the invention. Not even the patent holder at the university is entitled to use the invention if an exclusive license has been given to another. However, almost every organization will reserve for itself the right to use the invention for nonprofit research purposes. The biotech entrepreneur will always want to obtain an exclusive license on IP from which they plan on building a business.

Nonexclusive/Coexclusive Licenses

A nonexclusive license is a license granted to more than one entity for the use of the same IP within the same field or scope. For a biotechnology company, nonexclusive licenses to IP for core technology assets are not desirable. Nonexclusive licenses to process activities secondary to your product are common, such as those for polymerase chain reaction (PCR) usage in assays, or delivery methods for therapeutics. Coexclusive licenses are where there exist multiple applications of the same technology but licensed to different groups, operating in different fields. For example, under this type of arrangement, one licensee may use the technology in the field of human medical use, and another licensee may use the same technology in animal veterinary use. As a general rule, always seek exclusive licenses for the IP that is core to your technology and product.

Summary

The following three criteria should be carefully evaluated and found favorable before proceeding to license any technology. The quality of the science and the breadth of the market opportunity are significant assets and important criteria when raising capital for any new company. There are additional market assessments that impact success, and these are discussed in Chapter 7.

1. **Market Attractiveness:** Is there a market need large enough to generate potentially a billion dollars in revenue, or enough interest to build a profitable niche market? Does the product have an unmet medical need such that substitutes do not satisfy and will there be a demand for your product once is it produced?

2. **Scientific Merit:** Is the science and technology of significant quality with support in quality peer-reviewed journals and also support from key opinion leaders? Are the patents strong enough and have a FTO opinion? Will there be inventor involvement in this venture?

3. **Core Technology or Single Product:** Are there other applications or products that can arise from the same core technology? Is there a distinct advantage such that the first product development process accelerates the development of the second and third products in the pipeline.

Before licensing any technology, thoroughly evaluate it in light of the issues discussed before proceeding. There are many technology opportunities available, but the first step toward improving a company's opportunity for success is to start with the best core technology, capable of producing products with the best market opportunity. After a license is obtained and the company is making scientific progress, be sure to publish as much as possible in top-tier journals without jeopardizing the IP or product development lead. Doing this will significantly help in raising the much needed capital for product development, and improve the likelihood of attracting partnerships and alliances (discussed in Chapter 12). Also, be sure to continually identify other barriers-to-entry to keep competition at bay, including additional IP, and market and corporate advantages, which will further protect the licensed investment.

Chapter 4
Significance of the Right Business Model and Managing Risk

All companies operate within an underlying business model strategy. A business model is analogous to a frame upon which an architectural structure is built. This frame predetermines the height, width, and reach of the finished structure. If the frame does not reach a required height, or extend in a critical direction, the exterior construction cannot compensate for this limitation. A business model functions in much the same manner. Business models can be thought of as the collective way a company intends to make money. It is the sum total of all the strategic business approaches and their interrelationship to other parts of the business and the external world. It is critical to strategically position the technology opportunity into the appropriate business model prior to raising money for the company.

Business models can be extremely simple or highly complex. Developing a business model is a level of business strategy that is not typically reviewed by the vast majority of companies. The reason is because business models are usually firmly established and company employees are merely executing their roles. However, a development stage or start-up biotech company has the opportunity to identify a business model that best suits its technology and product needs while providing a competitive edge to sustain the organization in a changing market environment.

Business Model Examples

A comprehensive study of 10,970 publicly-traded companies across all industries was evaluated for six measures of performance.[1] Malone et al. categorized these companies into 16 types of business models, and they found that some business models performed better than others. Although some business models only work

---

[1] Do Some Business Models Perform Better than Others?
T.W. Malone, P. Weill, R.K. Lai, V.T. D'Urso, G. Herman, T.G. Apel, and S.L. Woerner. MIT Sloan School of Management, MIT Sloan Working Paper 4615-06, May 2006 (http://ssrn.com/abstract=920667)

for certain industries, many are reincarnations or slight modifications that are adaptable to multiple industries, including the biotech industry.

Let's begin with an easy-to-understand consumer business model found in several industries called the manufacturer-retailer model. In the computer industry, a company will manufacture personal computers containing the latest and greatest features. Manufacturers stock finished product and then ship them to retailers, who then inventory them and later sell them to the end-user customers. Retailers such as Best Buy, Comp USA, and even Wal-Mart purchase computers from manufacturers at wholesale prices, add a mark-up or profit margin, and then pay the manufacturer at the time of delivery or at the time the end-user purchases the product, depending upon where the strength of the relationship lies. Within the computer industry, Dell Computer pioneered an alternate business model that proved to be very successful in their early days. Dell established relationships via phone and Internet with end-user customers who placed orders directly with Dell. Because they only produced computers *after* the customer ordered them, they eliminated the retailer completely. The customer then received a custom-built computer with all the features they wanted without paying for unnecessary ones they did not want or need. Even better, the customer purchased computers for prices below that of Dell's competitors. Dell executed this business model strategy very well, but the underlying driver of the business success was that they found a business model that better met the needs of a targeted customer segment – home users and small businesses. A financial advantage of this business model was that Dell did not need to stock finished product inventory (in the computer chip industry, parts became obsolete quickly) nor did they need to sell products at wholesale prices to retailers, but rather sold at retail prices directly to the end-user. Also, because of their large volume and buying power, they successfully pushed their parts inventory to their suppliers for a just-in-time delivery, which significantly cut their operating costs and increased their net income. Using this alternate business model strategy, Dell grew rapidly and profitably. Dell was successful in spite of larger competitors in this industry because they selected a business model that better met their customer needs. Because of competitive forces and shifting markets, Dell's business model has evolved to include the manufacturer–retailer model in a portion of their business. In this example, Dell's business model more strategically targeted a segment of customers, provided better financials for the company, and better met a market need, resulting in a more satisfied customer.

Biotechnology Business Models

In the pharmaceutical industry, a venerable business model is the Fully Integrated Pharmaceutical Company (FIPCO), which is sometimes referred to as a "vertically integrated" company. In this model, a single company performs all the functions of the business – from initial drug discovery through to final marketing of their products. FIPCOs have the capability to perform research, development, clinical trials, regulatory, manufacturing, and marketing functions, with broad and vast expertise contained within the organization. Although this business model lets companies control all aspects of the product development through to the marketing – it comes at a tremendous cost. Pharmaceutical companies can typically afford to be "fully integrated" because there is a significant profit margin in pharmaceutical drugs. Unfortunately, drug discovery and development pipelines can quickly become sparse and companies end up with latent downstream capacities in the regulatory, clinical trials, manufacturing, and marketing departments. To be more competitive, the FIPCO business model has evolved in ways to continue generating the types of profits expected of pharmaceutical companies. Some of these modifications include reducing downstream capabilities and outsourcing some of the clinical trial functions to Contract Research Organizations. Instead of discovering all their drugs internally, FIPCOs are also augmenting their R&D capabilities by purchasing or in-licensing technology and drug leads from smaller biotechnology companies. By modifying their business model, they improve the efficiency of their organization. For most start-up biotechnology companies, becoming a FIPCO may be their ultimate goal but this is not a realistic or believable one – doing this job well is even difficult for mammoth pharmaceutical companies. Though it is possible a small biotech company could achieve this in the future, it is hardly a business model that would be taken seriously by any sophisticated investor. In fact, stating a FIPCO goal for a start-up would tell knowledgeable investors to run away – and fast. Good companies will constantly evolve their business model to provide the most strategic advantage. Even now, pharmaceutical companies are evolving to an alternate business model called FIPNet or fully-integrated pharmaceutical networks, where they believe they can improve productivity and increase efficiency by collaborating out parts of their functions.

The Drug Repositioning Business Model

To speed the drug development process, some biotech companies have adopted a drug repositioning business model that reduces the long R&D cycle and accelerates the drug safety component. A biotech drug repositioning company uses technology and know-how to identify an alternate or narrower population of users that can benefit from a shelved drug which previously passed safety testing in humans. This business model aim is to position abandoned drugs into alternate intended-use populations based upon the drug's mechanism-of-action, or by finding a population in which there is reduced side effects and greater efficacy. By adopting a drug repositioning business model, some biotech companies are able to shorten the drug development time and leapfrog into initial human safety testing. Gene Logic has a business model that partners with pharmaceutical companies to reposition abandoned drugs and share in the profits if the drug is brought to market for a different indication. In the drug repositioning business model, some companies focus on drugs that have completed Phase II clinical trials with demonstrated safety profiles in humans. These types of biotech companies take advantage of previously sunk development and clinical trial costs, which can save up to eight years and hundreds of millions of dollars in drug development. Because this accelerated pathway reduces risks and saves time, the biotech drug repositioning business model has been appealing to a certain segment of investors.

The drug repositioning business model has examples to point to for successful product precedent. Many drugs that failed for a particular indication were later found to have success for use in other indications. One notable repositioned drug that failed as an antihypertensive drug was Viagra (sildenafil). Although human clinical studies did not show the desired efficacy as a antihypertensive, the "side-effects" were noted by Pfizer which then repositioned it as a erectile dysfunction drug. Other examples of successfully repositioned drugs include Eli Lilly's cancer drug Gemzar (gemcitabine), originally developed for use as an antiviral; and Evista (raloxifene) originally developed as a birth control drug, then repositioned as an osteoporosis drug, and later found another indication as a prophylactic drug for the prevention of breast cancer. Biotech companies that adopt a drug repositioning business model have their own set of risks, but they exchange them to mitigate the later-stage risk of clinical safety.

Molecular Testing Business Models

A successful business model in the molecular testing industry is the development of molecular tests that run on proprietary platform instruments developed and manufactured by the same company. Molecular diagnostic companies that use this instrument/content business model successfully are Luminex, Becton Dickenson, and Roche. In this business model, the company will develop and market "content" (different tests) that only operate on their instruments, and do not run on a competitor's. Success occurs for this business model when they can install their platform testing equipment in to the largest laboratory market possible. In order for customers to have the desire to purchase these instruments, the company must also develop a desirable menu of molecular testing content that their customers want. The instrument/content business model is basically identical to the old business model of the razor handle and the razor blade. In the shaving industry, Gillette and Schick found they could almost give the razor handle away because they made most of their profits on the reoccurring purchase of razor blades fitted only to their razor handle. When a customer purchased a razor handle, the customer was committed. This is the same business model that the ink-jet printer industry uses. Ink-jet printer companies vie for consumers to purchase their printers, sometimes at a loss, because they make huge profits by perpetually selling their branded ink. In 2004, financial analysts note that although Hewlett-Packard's ink and toner business accounted for less than 25% of its revenue, it comprised more than 50% of HP's profits.

Another successful business model in the molecular testing industry is the performing of unique and proprietary tests as a service in their own clinical laboratory. This is the business model adopted by companies such as Myriad Genetics Laboratories, and Genomic Health. Myriad Genetics Laboratories has research, development, marketing, and sales functions for its molecular tests for predicting risk of hereditary-based cancers. In addition, they also maintain a clinical laboratory, which is certified and licensed to solely perform these laboratory testing services. Because their tests are not licensed to other clinical laboratories, and all testing is performed at the company's clinical laboratory, they retain all the profit. This is also a similar business model that is working successfully for Redwood City, CA based Genomic Health. It is conceivable that these two companies could also create an instrument platform that could run their genetic tests, and then sell these instruments to multiple laboratories under the instrument/content business model. However, their business model choice provides them with superior financial, market and regulatory advantages and allows them rapid modification and implementation of testing services.

These few examples of therapeutic and diagnostic business models demonstrate how their selection helps a company to be competitive with their particular technology or product. Although there are numerous business models used in the therapeutic, biologic, diagnostic, medical device, clinical laboratory, and research reagent industry, all are not adaptable to each sector. Be a student of successful business models, and understand why they are successful. Think about the strategies that would give a company a sustainable competitive advantage over the competition and lead the company to sustained profitability. Just as with the razor-handle/razor-blade model, portions of old business models may be applicable and transferable to new products or technology depending upon your needs and objectives. Study alternate business models in adjacent sectors and consider if portions of these may be adaptable to improve competitiveness or reduce business risks.

How Do You Determine the Best Business Model for a Technology?

Although we have only discussed a few business model examples, there are dozens that could apply to a biotech business. Sometimes the product opportunity is straightforward and the selection of a business model is clear. However, remember there are multiple portions of any business, and each portion can result in a hybrid of models. Do not assume that the technology will automatically dictate the best business model strategy. Start by first examining the unmet needs in your target market and then think about the best way your technology can help meet them through your choice of business model. The first step in defining the best business model strategy for a technology requires knowing answers to the following questions:

- How does the company intend to make money?
- What is the process used to develop and produce the product? Are there alternative ways to reduce the time, costs, or risks during development of this product?
- What is the nature of the relationship between the company, and its vendors and suppliers needed to produce the products or generate the services? Are there relationships that can be leveraged and are synergistic?
- Does your company have intermediaries in the value chain between itself and the end-user customers? Is there a way to leverage this relationship to improve the company's success?
- Are there any other external players in the value proposition to the end-customer to sell these products?
- What are the needs that the product will be satisfying for the target market? How is this accomplished? How does your company intend to reach its customers? How will the selling of the product be carried out and by whom?
- Does anyone else receive benefit when the company sells its product or services? Can these relationships be leveraged?

Each of these questions may not be applicable to all products; however, they are starting points to help think about the best way to leverage your technology and product opportunity into a market. By knowing the answers to these and other relevant questions, the best business model can be determined which will maximize the benefits of the company. Later, you can even refine the business model and tailor it to your product's particular strengths and market needs. Remember, a business model is dynamic and can evolve for various reasons, such as the availability of new market channels or changes that occur within the regulatory environment.

Becoming a Risk Manager

The best way to characterize the biotech entrepreneur's work in establishing and building a life science company is "managing risk." The careful selection of the company business model is the first step in managing the multiple risks, and maximizing the business success of the company. All choices in life carry a degree of risk and a level of consequence. When purchasing an automobile and choosing a particular make and model, or deciding whether to buy new or used, even these choices carry risks to its reliability, repair costs, and driving enjoyment. The level of risk for these choices is usually moderate, and the consequences of these decisions are usually minimal – unless one cannot afford to repair a lemon always in the shop. Choosing an interventional cardiologist to operate and perform a coronary angioplasty procedure entails a much higher degree of risk and a greater level of consequence then the previous example. Every decision requires a different risk assessment because there are different consequences for a poor choice. The choice of business model is critical and the proper selection will greatly reduce the future risk to the business by selecting the best one. Spend the time necessary to select the proper business model.

Why Risk Assessment and Risk Management Is Important

You must begin any business with a good appreciation of the business risks and address them as early as possible – this will greatly improve the odds of success. A great technology cannot succeed with poor management, and a great technology with great management still cannot be successful without a great market strategy. Still, a great technology, a great management team, an excellent market strategy cannot be successful without solid intellectual property protection and broad patent coverage. My point is, that it takes excellence in many criteria for a biotechnology company to succeed, and all these must be integrated. Having a good method of risk assessment for your company is fundamental to your ability to raise capital, and to know what you need to do to increase your likelihood of success. As we will discuss in Chapter 6, when you reduce commercialization risk, you increase the value of your company – learn to be a good risk manager.

What Are the Risks That Need to be Managed?

There are at least five risk criteria fundamental for success in the biotechnology industry, and these are listed below in the Biotech Evaluation Tool. Obviously, there are more than five risks factors, but one can examine most failed biotechnology start-up organizations and trace their problems back to one, or several, failures within these five risk criteria. Each of the risk criteria is exceedingly important since there are multiple ways to fail, but a company must be strong in each of the five areas to succeed. Possessing strengths in four of the five criteria does not compensate for a weakness in the fifth. The entrepreneur must integrate and manage all five risks to have the best opportunity for success. Integrative risk management is the ability to manage all risks simultaneously, and understanding how one risk impacts each of the other risks.

Use this tool and the accompanying questions to determine a ranking of your start-up business according to each criterion. I developed this list to quantify and score a company/business/technology using a 1–5 scale, with 1 being the worst and 5 being the best, when compared with the best in the industry. After evaluating each criterion independently, I collectively score them. If any of these five risk criteria are below an acceptable level (indicated by a score of 3 or below), this reveals an unacceptable level of business risk for an investor. When I evaluate a start-up company for investment advice, or analyze an organization's potential for success, I use a more detailed version of the following five criteria. When performing this evaluation I include other facets of the organization and management because they are intricately tied to future growth success; however, these are not critical for this particular exercise.

It is virtually certain that a start-up or development stage company will not score well in all of these categories, so do not become discouraged – it does not mean that the company is doomed. A low scoring biotech company at one point in time can be very attractive years later after certain risks have been reduced or overcome. It is important to use this tool critically to identify any company weaknesses, then direct your focus on what to do to improve as it is impossible to fix an unidentified weakness. Once areas of weakness are identified, draft a plan to overcome these, and include when and how it will happen. There will be plenty of opportunity in other situations to see the "glass half-full," but at this stage be very critical because it is certain that others will.

Assessing business strengths and weaknesses is absolutely necessary before beginning to raise capital. Potential investors evaluate a biotech business opportunity based upon similar questions, so I strongly advise using this method because it will help in preparing for answers to investor questions later. Revisit this tool periodically during the organization's development, and reference it often to evaluate improvements and changes. If a company is properly managed, these scores will improve over time. By demonstrating you are successfully managing each of these risks, investors will be comfortable with your ability to lead the organization's development future.

Biotech Evaluation Tool

1. **Management Leadership, Capability, and Past Success**
   (a) Is the CEO experienced in a leadership capacity, and what assurances demonstrate that he/she can guide this organization successfully?
   (b) How seasoned is the team, and do they have directly related expertise applicable to this company or industry?
   (c) What parts of their previous experiences make them suitable for a start-up environment, and what level of success have they previously accomplished that is applicable to this situation?
   (d) What collective capabilities does the company possess as a result of this management team?
   (e) How complete is the management team and where are the gaps. What plans are made to overcome this?
   (f) What are the weaknesses of this management team individually and as a whole?
   (g) Does the management team have the capabilities to fully execute the business strategy, and have they demonstrated this by previous experience?

2. **Technology Robustness, Applicability, and Scientific Team**
   (a) How innovative is this technology and its application relative to other new concepts and technologies in the industry?
   (b) What do the opinion leaders in this field have to say about this work and technology?
   (c) Does this organization have top scientific or technical staff that has demonstrated leadership in this field?
   (d) Has this team published proof-of-concept studies or other supporting information that has been accepted in top peer-reviewed journals?

(e) Has this scientific team won peer-reviewed grants (NIH, NSF, NCI, etc.) that indicate their work is respected by their peers in this field?

(f) How solid is the intellectual property protection? Is there a freedom-to-operate opinion?

(g) Is this a core technology or just one product?

(h) What are the potential future products from this technology?

3. Market Demand, Positioning, and Barriers to Entry

(a) Have they been able to prove that there is a demand in the market for their future product or service, or will they be creating a demand?

(b) Is the marketing strategy sound and based upon solid evidence supporting that there is an unmet market need for this product where substitutes do not suffice?

(c) Is the market large enough? Will it support the kinds of returns needed for the company to become a sustained success once the product is commercialized?

(d) Who are the competitors, and what are the product substitutes which could displace the market demand for the product?

(e) Is the market converging toward the future product need, or is it slowly moving away from where the product will be positioned?

(f) What is the planned distribution channel to reach this market? Does a distribution channel already exist or do they need to create a new one for this product to be successful?

(g) What are the Pro forma projections and how much of the market can they reasonably anticipate capturing with their product?

(h) Does the company have the personnel with the capabilities and know-how to reach this market?

4. Regulatory Hurdles

(a) Does the team know the process and the length of time estimated to obtain regulatory approval for the product or service?

(b) What are the risks for regulatory approval with this type of product? Are there any successful examples of others or are they creating a totally new category?

(c) Does the company have direct knowledge and communication with regulatory industry experts who are intimately familiar with the ongoing issues in the approving agency?

(d) What is the status of regulations for the product in the market the company is entering? Are the regulations clearly defined, or are they dynamic with future changes anticipated?

(e) Are there impending regulations that are as yet undefined?

(f) Does the company have a well-developed plan with the expertise in the regulatory strategy and process?

5. Financing Suitability

(a) What is the potential of the organization for continued funding based upon other similar types of organizations and funding trends?

(b) Is the business a good candidate for venture capital funding based upon venture capital funding trends?

(c) What is the exit strategy for the company? Is it attractive enough to show a potential 10× return on investment?

(d) How much cash is in hand, and how long with it last at the current rate of operations? What types of investors are currently invested (Angels, VC, Friends and Family)?

(e) What is the total amount of money estimated to reach the market and how much is needed when? Are there fundable milestones that can be reached to improve the chances of securing follow-on funding later?

(f) Are the outlined use-of-proceeds for the development of the organization reasonable throughout the product development cycle?

Summary

The proper selection of a business model must be based upon the collective assessment and integration of all risks in order to provide a sustainable competitive advantage with sustainable profits. Selecting the best business model improves the opportunity for success of a company, and this is the first step in risk management of the business. Know the business models of competitors in the industry, and understand what makes them strong. There are predictable strengths and weaknesses for each business model, so learn what they are before choosing an appropriate one. Become a student of business models and be familiar with the reasons one works and reasons the same business model fails in another situation.

As risk managers, the goal is to constantly manage and reduce the business risk for an organization. For an early stage business, there are many decisions that have long-term impact: determining your product development pathway, protecting your intellectual property, determining the method and timing of financing, deciding which individuals to hire, and choosing your marketing strategy. No one can be expected to completely eliminate risk. The objective is to know the risks, and make proper decisions to manage them. Realize that success does not come from just making right choices. Much is determined by creativity and initiative in developing ways to circumvent challenges rather than just making choices that provide the lowest risk.

Finally, understand the five risk categories of a business, and objectively list each strength and weakness. Develop a plan to bring these to a level of excellence. The Biotech Evaluation Tool should also be used as an aid in preparing to raise capital and to establish a reference point during the company's development to gauge future progress and improvement. The outline and answers to these questions will be helpful when creating a business plan as discussed in Chapter 9.

Chapter 5
Legally Establishing the Company

If you are like me, you may be a person who buys "self-assemble" furniture and starts putting it together while looking at the pictures as you take pieces from the box. Instructions? Those are for people who can't figure anything out by themselves! Although I can remember the sinking feeling, holding three extra bolts, trying to figure out if they were really "extra" or if I missed something! Unlike self-assemble furniture, a budding biotech business cannot be disassembled and reassembled. If you are the personality type who likes to figure things out as you go, and you did not read the previous chapters – go back to the beginning because it sets the foundation we build upon in subsequent chapters. Building a biotechnology company is challenging, and has been likened to assembling an aircraft while it taxis down the runway. You must feverishly work to complete it before it takes off. This analogy resonates with many experienced biotech entrepreneurs who have come desperately close to running out of runway.

The purpose of this chapter is to orient you to the legal and business matters in establishing a biotech company. We will also discuss any impact that these issues have on growth and attracting the type of capital needed to grow the business. For those with a deeper interest in biotechnology law matters, I have included references to a couple of good books on business and biotechnology law, to provide more detailed information on establishing a biotech business.

The First Step: Find a Good Attorney

Now that you have licensed a novel technology that leads to a biotech product for an unmet market need, and have identified a suitable business model, you need to formally establish your company. The next step is to find a good attorney. You would not take medical advice from your mechanic, so do not take legal advice from friends – find a good attorney to help you establish and counsel you in your business. You might ask "Why do I need an attorney, aren't there pre-printed legal forms available for most everything? Can't I just fill in the blanks myself and save a lot of money?" The answer to these questions is "yes." In fact, most attorneys use their own "boiler-plate" forms, and seldom create many new documents from

C.D. Shimasaki, *The Business of Bioscience: What Goes into Making a Biotechnology Product*, DOI 10.1007/978-1-4419-0064-7_5,
© 2009 American Association of Pharmaceutical Scientists

scratch. However, when hiring an attorney, you are paying for legal advice and business guidance based upon their extensive experience – not someone who fills in the blanks. You need the advice of an experienced attorney for things such as: the impact of terms for founder's agreements, implications of tax law, strategy for issuing stock options, best practices in intellectual property (IP) protection, interpreting employment law matters, and structuring various contracts and agreements. Also, as you prepare to raise money, you need legal guidance with securities issues and help in negotiating and understanding a venture capital term sheet. Choosing the right attorney is one of your most important early decisions. Think of the selection of your attorney as the hiring of the most critical employee for your organization. The choices you make in all matters of your business, such as financing and organizational direction, will be impacted by the counsel and advice of this individual.

How Many and What Types of Attorneys Do I Need?

For a biotech business, there are at least three types of legal expertise that you will need during your company formation and development. They include the following experiences:

Corporate and Business Matters: An attorney who specializes in biotech start-up organizations and practices business law

Patent Prosecution and Litigation Matters: An attorney who specializes in patent law and biotech patent prosecution and litigation

Securities and Private Placement Matters: An attorney who specializes in private placements and securities laws

These may be separate individuals, or sometimes attorneys will have combined experience. For instance, you may find a great corporate attorney with decades of experience in both start-up business and securities issues. However, for patent prosecution and litigation, you generally will not find someone extremely proficient in patent law who is willing and able to support other areas of law. As you grow and develop your organization, and as issues become more complicated, you will likely need separate individuals specializing in each of these areas. Later you may even have different attorneys working in patent prosecution and another in patent litigation, should you have to defend any of your patents against competitors.

Corporate Attorney

During the start-up stage, the first and most important attorney is a corporate counsel. It is essential to find an attorney that specializes in start-up issues such as organizational structure, employment agreements, issuing stock options, and financing structures – particularly venture capital deals. An ideal corporate attorney will be familiar with the typical start-up issues a company will face, and will give advice on other issues related to establishing the company. The best corporate attorney is someone who not only understands business law, but also has good business acumen and is able to advise on strategic business decisions. Ideally, this will be a long-term working relationship, so be sure that you work well together, and most importantly, find someone who is a good listener and can explain things in a way that is understood.

Patent Attorney

The second type of attorney needed is a patent counsel. This is an attorney who specializes in IP law, and should be someone who understands the scientific field in your specific technology area. If the IP was licensed from a research institution, the Technology Transfer or Licensing office of that institution usually retained the best patent counsel to handle their IP needs. Oftentimes a licensee can "inherit" a good patent attorney with substantial history in this technology. This person will intimately know the prosecution history of these patents. Many times, agreements can be worked out directly with this patent attorney, or in conjunction with the licensing institution. At one start-up company, we retained the same patent attorney that the research institution used for all patents we licensed. We then negotiated with the research institution to continue managing this patent portfolio and to share the filing and prosecution expenses over a period of time, deferring these payments until we reached a significant funding event. If you do not have confidence that the institution retained the best IP counsel, or if you prefer to work with someone else, seek a patent attorney with a combined background or dual degree in your field of interest such as a J.D. plus a Ph.D. or Chem.E. These types of individuals bring added-value because they can comprehend the science quickly, and add to the patent and claims in ways that only experienced scientists can. A good patent attorney will help develop the IP strategy to protect the products against future competition. Having a strong IP portfolio position is essential to attracting the type of investors needed at subsequent stages of the company development. Protecting IP is not an area of business to do on the cheap. Allocate a realistic amount of capital to this portion of the business, and be creative in finding ways to get the work done within a budget.

Securities Attorney

The third type of attorney needed is a Securities Attorney. A securities attorney provides guidance on issues related to raising capital, complying with securities laws and protecting the company interests as you raise money. Sometimes the corporate attorney may handle less complicated matters of securities law directly, or they may seek counsel from a securities attorney for more complex matters related to raising capital. Depending on the size of the law firm they may also have in-house securities attorneys who can help on issues related to raising money in a private placement, or preparing offering documents. Your corporate attorney should still be involved or informed about decisions made by any securities attorney; they will help through their knowledge of past issues and their understanding of the future direction of the company.

How Does One Find a Good Attorney?

It is important to find both a corporate attorney and IP attorney very early on during the establishment of a company. Most companies have limited capital to start with, and it is essential to have some creativity in working out financial arrangements. So how does one go about finding a good attorney who specializes in entrepreneurial life science companies? Start by asking other entrepreneurs who started biotech businesses, which attorneys they have they used, and who they would recommend. Find out why they recommend them, and in what ways they helped their organization. If you cannot find an attorney by recommendation, start by contacting reputable law firms and ask them if they have attorneys specializing in start-up biotechnology or life science companies. If you leave phone messages, observe how long it takes them to call you back – you will want someone who is responsive to your requests. Good attorneys will be busy, so do not necessarily expect an immediate response. The backlog for good corporate attorneys can be analogous to restaurant foyers. If the restaurant is good, their foyers always have people waiting. If an attorney does not get back to you within 24 hours, they may be too busy, or they may not be particularly interested in your business – so keep looking. Try to find a local attorney if possible, because when you need face-to-face meetings you do not want to always be getting on a plane. However, if you are starting your company in a non-biotech cluster or an area not well-established with technology start-ups, you may not have many options. In this case, search the internet and make a number of carefully selected calls to law firms in biotech cluster cities. There are compensating advantage to having a corporate attorney in a traditional biotech cluster. Good corporate attorneys in the biotech industry usually have venture capital contacts and access to seasoned biotech executives, so this could help with some of your future financing and recruiting needs. A good corporate attorney is someone who will grow with you throughout your company's development and their familiarity with your corporate history makes them effective and efficient with their advice and their fees.

For start-up companies, I generally do not recommend starting with mega law firms unless they have smaller practice segments within the firm, such as several attorneys specializing in start-up life science organizations. This is because extremely large law firms usually hire a large number of associates and junior staff, who are less experienced and may end up doing most of the work. You really do not want to be paying a junior attorney to be learning on your nickel, even if they are employed at the most prestigious law practice. Mega law firms have highly specialized attorneys usually in one particular area of law, so it may be more difficult to find an attorney who has good working knowledge in the breadth of your needs. For this reason, I prefer to start with the smaller to medium-sized law firms, and find a partner or senior member who will work directly with me. It is better to work with a full partner in a small to medium-sized law firm who can give the best and most experienced advice. Realize, however, that partner fees are significantly higher than those of an associate or junior staff member, but you get what you pay for – experienced advice.

Talk with several attorneys before making a decision on one. A good question to ask during an attorney interview is "Are you familiar with a Venture Capital Term Sheet and can you explain the meaning of some of the most significant terms?". There is much to be learned by this exercise. First, you will learn if they have patience with you and your remedial questions, and second, you will find out if they are a good educator and communicator. If you have a hard time understanding what they are talking about, or find they have difficulty explaining the substance of these terms, you may be in for a rocky relationship. It is also important that you and your attorney establish a good rapport, and share similar values in your approach to business. I do not mean that they need to become your best friend, but you must have confidence in their judgment to give you the best counsel. Regardless of whether it is a small, medium, or mega law firm, what you are looking for is breadth and depth of experience, a communication style that helps you understand the issues, and a relationship of trust based upon the sharing of similar values.

How Much Will This Cost?

All attorneys should give a complementary inital visit for you to discuss your situation and decide if you would like to utilize their services. If they insist on charging you for an initial consultation, you are better off finding another attorney. First, recognize that you will incur legal expenses greater than what you may have anticipated but getting a business established correctly will save you major headaches down the road. Cost estimates for legal help getting your company established and putting together documents can run from $5,000 to $25,000 or more. Start-up costs will depend on a variety of things, such as the complexity of the business, the number of founders and the number of issues related to a technology license. Costs for legal assistance to reach a closing on your first round of capital, including drawing up the documents, can run from $10,000 to $50,000 depending upon the complexities of various agreements, the size of the round, the number of investors, and the terms related to the funding. Attorney fees vary greatly depending upon locale and demand. Corporate attorney rates for start-up expertise in the biotech industry range anywhere from $200 to $750 per hour depending upon experience. Again, remember that you are paying for strategic legal and business advice so make sure that you will be getting what you need before engaging anyone, at any price.

When you need legal assistance, your attorney should give you a good estimate before starting, and some may charge you a flat rate for the work if it is clearly defined. For larger transactions, such as closing on a venture capital round of financing, you may be able to get a commitment for an estimate or not-to-exceed limit. Some attorneys that specialize in start-up organizations will take deferred compensation, but charge a higher fee and take a small equity position. If you know you have large upfront legal needs and are very short on capital, this is one option to consider. During our first Series A Preferred round, we negotiated with our attorney to defer some fees, and they had an option to convert a small portion into equity. Because of our good working relationship, they also discounted much of their services. Later, when we hit some rough financial times, they even allowed us to suspend payments for awhile until we reached the closing on another round of funding. At another company, our corporate counsel not only deferred payment, but also participated in a bridge loan to meet payroll during a period between financing. This aspect is yet another valuable reason for establishing a good working relationship with a partner of a law firm because they will have the authority to make these types of decisions.

Choosing a Company Name

A company name is something that you will live with for the entire life of your organization. Surprisingly, people may not think long enough about the perpetuity of a name. This sentiment may also apply to parents when selecting unique childrens' names like Dweezil and Yamma; however, there is great value in being unique – if you are successful, you are easily distinguished and recognizable.

Why Is a Company Name Important?

There are instances where the selection of a company name can be confusing to the public in both a positive and negative sense. Nanogen is a publicly-traded company located in San Diego, and started in 1993 having an early focus on development of platform technologies and medical diagnostics to run on their platform instruments. During the late 1990s and early 2000s, nanotechnology was receiving significant investor attention as a new technology wave. "Nano" meant the development of nano-sized particles that result in everything from better medicines, new delivery mechanisms, to new material sciences and better cosmetics. Nanotechnology companies began receiving significant boosts in their valuations and share prices – including Nanogen, even though they were not a nanotechnology company. This is an example of how serendipitously a name association can accidentally help. Be aware that this can go both ways. In the late 1990s and the early 2000s, any company that had a "dot-com" after its name was raising incredible amounts of capital. However, the subsequent "dot-bomb" era cleared the field of most all of them, especially those without a good business model because they were unable to generate any meaningful revenue. Subsequently, any company with a dot-com reference was avoided because of past failures in the industry.

You need to select a name prior to incorporation. It is important to have a name that somehow brands the company and its future. As the life science industry continues to expand, unique names for biotechnology businesses become more difficult to find. Sometimes entrepreneurs do not take enough time or give sufficient thought to the selection of a name and the issues that are related to its choice. Other entrepreneurs may anguish over this, such that it consumes too much time and energy from other critical matters. Just be sure to spend the time necessary to arrive at an appropriate name that reflects the organization. There are at least four aspects to consider when choosing a company name:

1. Does it somehow represent or characterize the current and future focus of the organization?
2. Is it relatively easy to pronounce and recognize?
3. Is it unique enough that it will not be confused with the names of other organizations?
4. Will the company name work well with product names envisioned in the future?

There are helpful ways to arrive at ideas for organization names. You can ask associates or employees for name suggestions or offer the person that comes up with the chosen name $100 or a dinner for two at a favorite local eatery. Another creative way to arrive at good names is to use the same guided brain-storming process professionals use in the advertising world. If you have a friend or colleague in this business, tap into their expertise or pick up a book that describes this process and follow it yourself. There are also services you can hire that produce computer-generated names based upon input you give them.

With a list of names in mind, place them in order of interest, then type them into an internet search engine and see what hits are returned. If thousands of hits are returned on each one, go back to the drawing board. It is not essential that you retrieve zero hits on a particular name to use it, but it is important not to get hits on names for other companies or products. Once satisfied that a name is not associated with another company or their products, provide these to your corporate attorney, who will then have an official search performed prior to usage and final adoption.

Can a Company Name be Changed?

There are certain situations where one may want to consider changing an established company name:

1. If the company has a troubled past that haunts the new management as they try to raise money, or you are reorganizing the company or doing a restart.
2. If the name is a source of confusion because it was strongly associated with a former focus and now the company has a new focus.

3. If the previous management had a notorious reputation and the new management wants to make a clear separation.
4. If the current name is problematic for business because it ties the company to an unrelated field.

Be sure to weigh the cost/benefit to the organization and its impact on any established efforts before making a company name change. However, for a relatively new company, there may not be much cost impact in doing this.

Incorporating the New Company

The next step is to incorporate the company and set up the legal structure. This is important for a number of reasons:

- It will reduce exposure to liabilities
- It will protect personal assets
- It will allow issuance of stock to founders/employees
- Investors will require this prior to investing
- It provides maximum advantage of tax laws including carry-forward losses for the business

There are five options for structure, and the selection of one is a discussion to have with your company attorney. The choice of structure will be based upon your current plans and future direction. These structures include the folowing:

- Sole proprietorship
- Partnership
- Limited liability corporation (LLC)
- S-corporation (S-corp)
- C corporation (C-corp)

Some organizations may start out as a LLC, until they get significant investments. However, because we are talking about a biotech company, ultimately it will need to be a C-Corp, which is the standard business entity for the industry. The selection of legal structure impacts how the business is taxed, along with differences in liabilities to the owners and fiduciary agents of the company. Your corporate attorney will advise you on the proper legal structure for starting the business.

When incorporating the business, your attorney will be filing Articles of Incorporation and Bylaws that the company will follow. This filing includes designating the number of authorized company shares, the number of Board Members, and other related issues. The state in which the company will be incorporated may be the state where it is located. If the corporate attorney is local, they will be most familiar with the corporate laws of that state. However, realize that prior to, or upon securing venture or institutional capital, most likely the company will need to be reincorporated in the state of Delaware where the corporate and tax laws are more favorable to companies. There is a cost associated with reincorporation and possibly a few other issues, but this is a discussion to have with your corporate attorney.

For those interested in learning more about these legal structures, I have included some good references. One of my favorite courses in business school was "Business Law for Entrepreneurs," taught by a practicing attorney who worked exclusively with start-up companies mostly in the life science area. The text was "The Entrepreneur's Guide to Business Law"[1], and I highly recommend this book as a resource for further interest when working through legal issues.

Ownership: Issuing Stock

The next thing your corporate counsel will assist with is issuing stock or stock options to the founders, inventors, IP holders, and key staff of the organization. Stock provides the stakeholders with equity ownership in the organization, from which they can benefit as they build value and when these shares can be sold. It is common practice to issue a certain percentage of the stock to the founders and management and then reserve a block of shares for other investors. Once the organization is established, it is beneficial to issue stock immediately, rather than waiting until a large amount of capital is raised. When shares are issued early upon formation, the founders can be given shares at a fraction of a penny per share. If stock is issued long after raising a significant amount of capital, there is a value imputed to the organization, and any shares issued at a discount to that price could have tax consequences. For instance, upon securing investor financing, someone has paid a specific amount of money for a portion of the company equity, so there is a "fair market" value imputed to these shares. The fair market value is the value that is fair in an open market, which is easily established by the price a willing buyer paid for the shares. If shares are then discounted to founders or key employees, there could be a tax liability based on the difference between the fair-market value and the amount of money these founders paid for them. For example, if shares are issued after raising $250,000 and the new investor received 25% of the total equity, the value of the organization is implied at $1,000,000 ($250,000/0.25 = $1,000,000). If shares are contemporaneously issued to founders for 25% of the organization but they did not pay $250,000 for these shares, they may have incurred a tax liability on a $250,000 gain. There is no reason for founders or key employees to be paying taxes on shares at this stage of the company development. This is just an example, but your corporate attorney will guide you through any tax consequences of issuing stock, or they will obtain the help of tax counsel.

---

[1]Constance E. Bagley and Craig E. Dauchy, 2003. "The Entrepreneur's Guide to Business Law" 2nd Edition, Thomson, South-Western West, pp. 730.

What Types of Stock Should be Issued?

Your corporate attorney will give advice on what type of stock should be issued, based upon the organization's plans and future capital needs. There are several types of stock that can be issued. These include: Founders Stock, Restricted Stock, Preferred Shares, Common Shares, Voting and Non-Voting shares, and two kinds of stock options: *Incentive Stock Options (ISOs)* and *Non Qualified Options (NQOs)*. These all have differing privileges and rights associated with them, and differing restrictions. ISOs are options typically reserved for existing employees, and have a different tax treatment then NQOs. ISOs usually can be exercised without tax consequences, and held without incurring a tax liability. NQOs are usually granted to nonemployees such as consultants and SAB members. NQOs are usually taxable upon exercise based on the potential gain, which is the difference between the exercise price and the fair market value, even if there is not a market for the securities. Although, with both these options, the owners may not want to exercise their rights until there is a market for these shares. There are other considerations with NQOs and ISOs, and your corporate attorney will provide counsel on these issues.

Restricted Stock is actual shares of stock granted to an employee. The restriction is usually the period of time that must pass before the employee actually owns the shares or can sell them. Grant sizes for restricted stock are usually smaller than grants for stock options to employees. The size of the grants for restricted stock can vary from about 25 to 50% to that of a stock option grant. This reduced size is because restricted stock represents actual shares given to the employee, whereas stock options are rights given to purchase shares later at a specific price. There also can be other types of restrictions placed on these shares based on how the grant is structured.

Vesting Schedules are usually given with stock options (NQOs and ISOs) and restricted stock. A vesting schedule is the period of time in which a portion of the stock option or grant becomes exercisable by the individual. Both stock options and restricted stock vesting are usually spread out over a defined period of time to ensure that all the founders and key individuals have incentives to remain with the organization. The typical vesting schedule is three or four years, where the first 25% may be granted immediately, and at the end of the subsequent year another 25% is granted (Cliff Vesting), then each month thereafter a portion of the remaining options is vested incrementally until the entire grant is vested at the end of the term. It is not unusual to issue additional grants of restricted stock or stock options containing new vesting schedules over the employee's tenure with the organization. Sometimes stock options and grants are referred to as "golden handcuffs," because they provide longer-term incentives for individuals to stay with the organization through tough times.

Each stock option grant or restricted stock grant will have a *Strike Price* or issue price, which is usually the *Fair Market Value* at the time the grant is made. There are legal restrictions and tax implications for assigning anything other than the Fair Market Value of the shares at the time of the grant. Technology companies typically make use of stock options as alternate means of compensation for employees. Some of these companies have gotten into trouble with the Securities and Exchange

Commission (SEC) for issuing stock options at historic dates where the Fair Market Value of the options were much lower than the price on the date at which the options were actually issued. Though this may seem like a generous thing to do for employees, it is in violation of SEC laws. This is why having a good attorney at the formation of your company is essential to help you avoid problems later.

What About the Founders?

Most biotech companies will begin with more than one founder. A benefit of being a founder is that they usually receive Founders Shares. If the business is started with several "Founders," the temptation will be to equally divide allotted shares among all. Your Corporate Attorney will help here, but first assess several things about each of the founding team members before issuing shares:

1. What has each contributed to the establishment of the organization?
2. Will their roles be equal in contribution to the new organization?
3. Will they all be working full-time for the organization?
4. Are they all committed to staying with the organization to see it through to success?

The answer to each of these questions will help determine the split of founders' shares allocated to each founder. Even though all founders may intend to stay with the organization to the end, when things get tough not all individuals continue to stay. Also, some founders may not want to work at this full-time. Circumstances may change in their lives regardless of their initial desire, and they may no longer continue with the organization after a period of time. For these reasons, it is important to have a *Founders' Agreement* that outlines the provisions and considerations given to the Founders in exchange for their work, contribution, and IP rights. Within the Founders' Agreement there should be a *Founders' Shares Buyback Agreement*. The Buyback portion of the agreement is a provision that the company can buy back a certain amount of the shares should one of these founders leave the organization in the future. The reason for this is that you would not want someone to walk away with a large percentage of the company who no longer contributes to the success of the organization. Should one founder decide to leave, it would not be reasonable to expect the remaining founders to carry the burden of creating value in the company and permit the leaving founder to walk away with a "free-ride" on the backs of the remaining founders. In addition, if any key employee leaves, some of that equity is needed to provide incentives to hire a replacement for that key position.

What Agreements are Needed?

The organization should have in place several agreements between its founders and employees. These agreements protect the IP assets and provide assurances that are important for any new investors.

A *Confidential Disclosure Agreement (CDA)* or *Nondisclosure Agreement (NDA)* is an essential document each individual must sign when hired with the organization. This document protects the company by requiring that the employee appropriately handle confidential information, with the understanding that there are penalties for disclosure of this information under certain conditions. By doing this, the company protects its know-how and IP from competitors.

An *Invention Assignment Agreement* is an agreement that may be combined together with CDAs or NDAs that transfer assignment of any and all new inventions conceived by the employee to the company. This ensures that the organization owns the IP it needs to develop and market its products. There are allowances given for previous inventions that the employee already developed prior to their hire. Be sure that all new inventions developed by any employee become IP of the company.

A *Noncompete Agreement* is an agreement to prevent an employee from quitting and starting an identical business in the same field, using the same technology. Noncompete Agreements are used to protect the company from founders or key employees going out and starting a competitive business with the same information that they have been using during the starting and development of this business. Noncompete Agreements have not historically held up well in courts, but your attorney will have some means to help protect an unfortunate or unlikely occurrence such as this from impacting the future of the organization.

An *Employment Agreement* can be combined with a NDA or CDA, and an Invention Assignment Agreement along with other provisions that constitute employment. This is a document that all employees should sign upon hiring, as delaying the execution of these types of documents can be problematic for upholding them. There are various things to consider for differing levels of employees, and your attorney will give guidance on who should sign, and what should be in an employment agreement.

Setting Up a Board of Directors

Your Articles of Incorporation will stipulate that a Board of Directors be constituted. A Board of Directors is a governance group that is ultimately responsible for the organization. The Board of Directors has a legal obligation to the company, in that they possess a fiduciary (trustee) responsibility to look after the best interests of the company, and only make decisions that are in the best interest of the organization. Once the company is incorporated, the Board of Directors is elected by the shareholders. At the start-up stage, a company may not have many shareholders besides the founder and possibly a few start-up investors. Carefully select the individuals who will comprise the Board of Directors based on their expertise and ability to contribute to the organization. Do not put friends and family on a Board unless they really are qualified to be Board members. Even then, be aware of the pitfalls of placing friends and family on the Board because difficult issues are decided by the Board and these relationships may influence their decisions.

Importance of Directors and Officers Insurance

The Board of Directors has a legal responsibility for the actions of the company, and have accountability to shareholders for their decisions. Because of this, it is imperative to have a Directors and Officers (D&O) insurance policy in place. This insurance helps protect the directors and officers in the event of a shareholder lawsuit that may result in judgment against the Board of Directors. There are various types of D&O insurance and the coverage amounts vary. This is another matter to consult with your attorney and insurance broker when constituting a Board of Directors. Realize though that no amount of insurance will protect Directors and Officers if there is willful negligence or abdication of their responsibilities. For this reason, it is important to carefully select a Board that will help make sound decisions about the future direction of the company.

Responsibilities of Your Board of Directors

Your Board of Directors has fiduciary duties and legal responsibilities to the company. These are summarized below:

1. **The Duty of Care** – the obligation to make decisions in a reasonable, careful, and prudent manner. All decisions have inherent risks, and any decision can be second guessed after events happen, which may bring into question the original choice. However, to attribute liability to the Board, someone must show that the Board members were negligent or acted in bad faith when making these decisions. The *Business Judgment Rule* is used when evaluating these decisions, which means that if the Board member made a reasonable decision that would have been prudent to make at the time, and used reasonable judgment and care, they have operated under the duty-of-care. The Business Judgment Rule protects the Directors from liability as long as they act responsibly.

2. **The Duty of Loyalty** – this means that all decisions or transactions with and for the company must not be motivated by self-dealing or any conflict-of-interest. It would not be unusual for a Board member, from time to time, to have a conflict-of-interest in a particular decision that requires Board action. The proper procedure here would be to disclose the conflict to the other members and abstain from voting on that particular issue. The Duty of Loyalty means that a Board member must not make any decision favoring self-interest, or that favors the best interest of anyone other than the company.

For most start-up companies, the composition of the Board of Directors evolves over the life of the organization, and the composition and expertise of the Board changes over time. During the formative stage of the company, there may only be three, or at the most five members on the Board including the founder or CEO. Usually the early Board members are individuals with a vested financial interest in seeing the organization become established, or those who want to help the company reach its first major funding event. If the technology came from a research institute or academic institution, and there was a strong financial tie other than just licensing, potentially a Technology Transfer officer may be on the Board. If early stage funds were secured from a professional investor with interests in this field, they may take a seat on the Board. The CEO, entrepreneur or founder would also be a member of the Board. If there are cofounders, usually only one founder will be on the Board, and they should be the best suited for this responsibility, not just that they were a cofounder. Select your Board of Directors based on their expertise, experience, and ability to contribute to the needs of the organization. However, after raising significant amounts of capital, Board member selection is usually tied to these large investments, and funding usually comes with rights to appoint a certain number of Board member seats.

A Board of Directors will have a Chairman. The Chairman's responsibility is to call the meetings and preside over the Board meetings. If the CEO of the company is not the Chairman, usually one of the Board members appointed by the major shareholders will be the Chairman. There are various reasons why the CEO should, or should not, be the Chairman of the Board. Some arguments against the CEO being the Chairman are that there is a concentration of power and less accountability. However, this depends upon the individual and their capabilities, and whether or not there is anyone on the Board more qualified to be the Chairman. At the start-up stage, this is not as much of an issue. If an outside CEO is brought in to take over or lead a start-up organization which was backed by venture capital, he/she will probably not have much of a say in who is on the Board or who is Chairman.

At some point in the development stage, the company will have raised its first or second round of institutional capital – most likely venture capital. A VC will require Board representation. Depending on the size of the investment and the number of VCs in the syndicate, the lead investor and other coinvestors may take a Board seat. If you are fortunate to have good venture capitalists with depth of experience in your field they will guide and strengthen the remaining Board member selection. As the company develops, there will be reasons to add additional Board members. The number can vary, but generally no more than five members are useful at early stages. Depending on the needs, independent Board members may also be on the Board. An independent member is someone who does not have financial or employment interest in the organization and can make decisions independent of any of these ties.

During late development stages of drugs or at marketing stages for diagnostics, there will be reasons to add other members to Board positions. One reason is because the company is preparing for a public offering. In this situation, the company needs to find individuals that have national reputations as experts in their field, or someone who has been successful with building a nationally known company. Sometimes companies may even seek retired heads-of-state or well known former public officials. These individuals can provide contacts, access, or credibility with certain facets of business and government and they may be a helpful addition on a Board. However, if this is just "window dressing" and these individuals are not really supportive of the vision and the management, but are more interested in their remuneration, no matter what their reputation, alternate individuals should be sought. Having well-known individuals on a Board can be helpful but they must be able to provide some valuable expertise that is not available from any of the other members. Adding additional members may increase your Board number to seven, although there is not a real good reason to have more than seven members on a Board at mid stages in the company development. Odd numbers of Board members are chosen for reasons of voting, especially when there are differences of opinions on direction and when there is no consensus. However, always work to get unanimous consensus from all Board members before taking any contentious path for the company. The support of the entire Board is always needed to work through the challenges of building any company.

How Should a Board of Directors be Compensated?

There is great variation in compensation for directors depending on the development stage of the organization and whether the directors are "insiders" or investors, and whether there are any independent Board members. Companies with Board members composed of investors and the CEO are not usually compensated additionally for their participation as Board members since they are simply managing their investment. As the company grows and independent Board members are added, their compensation is usually a mix of cash, such as an annual retainer and some form of equity compensation. Depending on the stage of the company, the cash compensation may simply be reimbursement for out-of-pocket expenses, or up to several thousand dollars annually. Generally, equity compensation for directors is given in the form of stock options, though it can also be in the form or other stock compensation as discussed previously. The options are usually vested over a number of years, typically four, where the directors can "earn out" their contribution and align their interests with that of the company. The amount of stock compensation may range between 0.25% and 2.00% of outstanding shares or more, depending on the value of these members to the organization.

Setting Up a Scientific Advisory Board

Who Are They and What Do They Do?

A key group of advisors important to the company's success is a Scientific Advisory Board (SAB). This is a group of individuals called upon for advice and assistance in matters pertaining to the science. Often the SAB is identified and selected even before the company is legally established because the expertise of some of these individuals may have been sought prior to licensing the technology. A SAB should be formed early on in the development of the company. Select individuals based upon their expertise and knowledge in the technology or science that the company is working in. These individuals should be considered experts by their peers. Typically, the inventors of the technology will be part of this group if they are not already employees of the company. Be sure any SAB member selected is truly capable of providing expert feedback, guidance, ideas, critical evaluation, and an expert opinion to guide decisions during the product development and even clinical testing phases of the company. Do not make the mistake of selecting individuals whose sole qualification is that they are investors but know little about the science or the business.

An SAB is not a legally constituted Board, and they do not have fiduciary responsibilities as the Board of Directors. For that matter, this group can be called a Scientific Advisory Committee if preferred, though traditionally they are collectively termed a SAB. The number of SAB members will vary. It is possible to have any number of SAB members, but three to seven is usually sufficient as long as they have sufficient time to participate. Have your corporate attorney provide a well thought-out *Scientific Advisory Board Agreement*, which contains the SAB member duties, the type of compensation, a CDA or NDA provision, and an agreement about publications and inventions.

A secondary purpose of the SAB is to bolster credibility for the science and to professionally benefit from the reputation of these individuals. Individuals considered "experts" in your field of science when associated with your work indirectly gives credibility to the business venture, and is reassuring to potential investors. Sometimes this purpose goes too far afoul, and in its worst incarnation it becomes "window-dressing" as just mentioned for Board members. The purpose of the SAB is not to window-dress for the public or investors, but to provide real input, help and advice in the development of the technology. Window-dressing can be easily spotted by someone looking into the involvement of these experts and by talking with them about their knowledge of the ongoing work. Window-dressing quickly becomes a deterrent rather than a help to the company, so select members who can really be utilized for help and advice.

A third function of the SAB members is to have them speak and present results at meetings on the scientific progress. There is credibility conferred when a member of the SAB presents at a national meeting on the data generated from your research or clinical studies, provided they played a role and understand it well enough to answer questions. Utilizing the SAB in this manner can speed acceptance of the company's work in the eyes of potential investors, as the SAB members will have credible reputations in their field. Having an SAB coauthor any peer-reviewed publications is also an important advantage and benefit, and shows their involvement and contribution in the development of the science toward a medically useful product. Just be sure that they actually contribute at an appropriate level before utilizing them in any of these capacities. When a company conducts excellent science, SAB members will enjoy participating if they are kept informed and called upon for their expertise.

How Often Should the SAB Meet?

The frequency of your SAB meetings will vary greatly depending upon the development stage of the organization and the technology. During the early stage, when a company is newly formed, the SAB may meet frequently, possibly weekly or monthly, though the full SAB may not yet be constituted. Once the full SAB is formed and selected, it would not be unreasonable for them to meet quarterly, but no less than semiannually. At later stages of product development when ongoing results or clinical testing results requires more time, or during marketing stages where the full SAB may not have much ongoing input, meeting less frequently, such as annually, is not unreasonable. It is possible to have more frequent interactions with one or two SAB members because they may have a particular expertise that is needed more than the others. However, do not forget to keep all members updated and fully informed on the scientific progress.

How Should the SAB be Compensated?

Since you recruit and retain the SAB early in your organization's development, it is usually understood that cash compensation is not generally possible. Like the Board of Directors, the SAB is typically compensated with either stock options or restricted stock in the company. The amount of stock options granted to SAB members varies depending on the company and the critical need for the individuals on the SAB. Ranges for stock options can include 0.1–2% of outstanding shares. Ranges for restricted stock can range from 0.1% to 0.5% of outstanding shares. If your members are highly sought after, sometimes you may need to pay a "per meeting" fee or nominal annual retainer to the SAB at early stages. However, it is not unusual to just provide equity and cover their out-of-pocket expenses to be present at SAB meetings. After later stage funding, you may add an annual retainer or per meeting fee when the finances of the company can support this. Stock options are a form of delayed compensation, and provide another way to ensure long-term interest in what you are doing. Individuals who are just looking for added cash compensation may not fully believe in the vision and purpose of your organization, or they may not be fully aligned with your goals.

A Clinical Advisory Board

At some point in your company's development, it may be helpful to form a separate Clinical Advisory Board composed of physicians who practice in the area of medicine your product will serve. Frequently, a SAB is constituted by both M.D.s practicing clinical medicine and also Ph.D.s conducting basic research. If the product is in molecular biology, bioinformatics, or genomics research, clinicians may have limited ability to contribute too specifically to your research, such as critiquing experiments, reviewing manuscripts, or assisting with writing grants. Whereas, when help is needed with clinical applications, clinical utility, and medical practice applications, the basic research members may not be able to provide much advice. Having a Clinical Advisory Board can be of great help for insight into medical adoption of a product whether it will be a medical device, molecular test, or a drug or biologic. All biotechnology companies do not have a Clinical Advisory Board. However, I have seen this group work well for later stage companies where clinical advice is needed and clinical experts are essential to the business and market acceptance.

The Virtual Company

At this point, you have found a great corporate attorney, legally established your business, identified a great Board of Directors and SAB, and issued your stock, but you may not have a physical address for your business. Having a virtual company is essential because early stage companies do not have large amounts of capital to devote to brick-and-mortar buildings. You may, or may not, even have an office to work from, or you may be sharing time at your current place of business. Be sure to discuss any arrangements like this with your corporate attorney as there can be potential issues if you do not work out arrangements with current employers when using their space or equipment for your new venture. We discuss more on the idea of a virtual company in Chapter 11.

Summary

Although I have said this multiple times, it bears repeating – find a good attorney with expertise in starting up life science organizations because their advice and counsel is critical in becoming properly established. They will help determine the timing and importance of various agreements and help firmly establish the legal foundation of the company. If you have selected a corporate attorney but later find that you do not have confidence in them or their work, you would be better off seeking the services of another you trust before going too far down the development path. You don't want to learn later that the optimal route may not have been taken for the organization's development, or that critical agreements were not drafted appropriately. When working to attract investors, you do not want them to be turned away by a poor organizational structure, dysfunctional agreements, or a mass of mistakes they do not want to fix.

Your corporate attorney will help set up an equity component that provides the best incentives and future compensation for the founders, management, and employees, to retain them while progressing through future development stages. Be sure to select Board members and SAB members carefully, and communicate with them often, seeking their input and advice. These are individuals to rely upon to help shoulder some of the challenges throughout the growth of the organization.

Chapter 6
The Product Development Pathway: Charting the Right Course

Development of a product is the key focus of all biotechnology companies. Detailed and accurate planning of the development pathway is vital for success. In this chapter, we review a prototypical product development pathway and identify some examples of things that must be accomplished during each stage. Product development is so diverse that it is unrealistic to characterize all pathways as similar; however, this may serve as a helpful example. All product development stages of a company need to be accurately characterized according to their particular product. If these stages are not well-defined, it is impossible to correctly estimate the amount of capital required to support product development through to each successive stage. If the company fails to reach the next product development stage, it generates unfulfilled expectations, making it more difficult to raise the next round of capital. Companies are rewarded for meeting goals, not for running out of money. By reviewing some general aspects of product development for therapeutics and diagnostics, the reader should be better able to outline their product development plan, and then estimate the amount of capital required to complete each stage.

It Is About Creating Value

A start-up company possesses little or no financial or fixed assets – yet it has the potential to be a billion-dollar company. How is that possible? Throughout the product development process, value is created. Value is rarely created proportional to time or money invested. Rather, value is created incrementally each time a subsequent product development stage is reached. Completing certain product development stages confer more value than others and with each successful step, financial risk is simultaneously reduced – value increases as risk is reduced. Steady product development progress in a financially efficient manner creates a situation where more investors will want to participate in this endeavor.

Throughout this book I make reference to investors, and their expectations within the context of most subjects discussed – including this one about product development. This is because everything done by the company must be viewed in such a way that

C.D. Shimasaki, *The Business of Bioscience: What Goes into Making a Biotechnology Product*, DOI 10.1007/978-1-4419-0064-7_6,
© 2009 American Association of Pharmaceutical Scientists

it does not detract, but rather increases the likelihood of raising additional capital. This is not to imply that a company will be forced to do things in a less desirable manner simply because it needs investor funding. It just means there is not the luxury of making many mistakes, or doing things haphazardly if a company hopes to be successful.

Biotechnology R&D: Little R, Big D

Academic or basic research is different from research in the biotech industry. The goal of academic research is to better understand the biology and mechanism of how and why biological processes occur. The goal of research in a biotechnology company is to use this research to develop products. These objectives are vastly different, yet they are based upon the same science. For this reason, research in the biotech industry is sometimes referred to as "Translational Research." Translational research can be thought of as systematic experimentation for the sole purpose of translating scientific discoveries into useful products or tools. Because the biotech industry is composed of many former academic researchers, sometimes it is easy for research scientists to migrate back to conducting basic research. Research scientists can sometimes succumb to pursuing interesting experimental paths that may not always advance the development of the product. Without lengthy industry experience, it is not easy to differentiate between early experiments that may be tangential to the final development of a product. This is another reason why it is important to have a clear product development plan that will help keep the R&D on course.

A Sense of Urgency Down an Unfamiliar Path

It is essential to have a clear development plan. It is also essential for the R&D team to have a sense of urgency in their work, because biotechnology product development always takes longer than originally planned. Unexpected technical challenges appear. Often, unplanned time must be spent to first solve a new scientific issue before getting back to the planned path for a product. At times, developing a biotechnology product may seem like "organized chaos" even in spite of well thought-out plans.

If you have ever rented a vehicle while traveling in an unfamiliar city, you may have experienced the underlying feeling of "lost-ness" and disorientation to your surroundings. Even with maps and detailed instructions, this feeling is similar to the one you may have when launching a biotech business and traveling down the product development path. As you navigate past the guard gate onto the freeway, you glance into your rear view mirror and see local resident drivers (these can be likened to your investors) impatiently waiting as you search for control knobs on the dash, and make a hasty decision where to turn. As you locate your turn signal, you quickly glance to your misaligned side-view mirror and see the words *"CAUTION: objects are closer than they appear"*. This is the same admonition to heed when navigating the product development pathway. Everything happens fast – money does not last as long as anticipated – obstacles *are* closer than they appear. By familiarizing yourself with product development obstacles you can mitigate their surprises. Anticipate obstacles, but do not let them stop progress. Be persistent and you can transform these apparent roadblocks into accomplishments. Always keep the focus on where you are headed regardless of what obstacles are in your path. Remember Henry Ford's words, "obstacles are those frightful things you see when you take your eyes off your goal." The successful companies are those that press on in spite of the challenges, and do so with a sense of urgency.

Two Types of Product Development Categories

Biotechnology products are diverse. They include products such as therapeutics, biologics, medical devices, laboratory tests, diagnostics, enabling tools and vaccines. The development path and time requirements for each will vary greatly. Because of this, it is impossible to address each development pathway individually. However, in order to outline general product development stages, we will cluster all biotechnology products into two product development categories termed "Treatments" and "Analytics and Tools." Though there are product development variances even within these categories, most development stages are similar enough that some will find application to their particular product. This outline serves only to provide a general aid for what to expect when developing a product. Also, know that some of these stages overlap, still, defining them helps to separate and measure progress.

1. **Treatments: Therapeutics, Biologics and Vaccines** (including drug delivery and gene therapy). This category encompass all traditional small molecule therapeutics, recombinant DNA produced drugs, and biologics. Examples of these are human insulin, tPA, erythropoietin and recombinant human growth hormone, monoclonal antibodies such as Synagis for Respiratory Syncytial Virus and Hercepin for treatment of breast cancer. This category also includes vaccines, gene therapy, and any molecule used to treat a disease or condition.

2. **Analytics and Tools: Medical Devices, Laboratory Tests, Diagnostics and Enabling Tools** (including platform instrumentation for both research use and commercial testing, clinical laboratory testing and all types of medical devices). *Analytics and Tools* refer to *In-Vitro* Diagnostic Tests, such as rapid point-of-care tests for HIV and other infectious diseases. It also includes genetic tests that are run in a clinical laboratory as a testing service such as genetic tests for gene expression and personalized medicine testing. This category also includes medical devices that support or improve human function, such as cardiac pacemakers, implantable insulin pumps, new medical imaging technologies or platform assay instruments used by research or clinical laboratories. With the convergence of technologies, medical devices now can include combined drug/devices such as drug-eluting stents. These medical devices have different development and regulatory challenges, but because they involve the use of drugs they are an amalgam of both Treatments, and Analytics and Tools.

Treatments can require 8 to 15 years to move from discovery to commercialization, whereas Analytics and Tools can reach the market much quicker, generally from 3 to 7 years based on the complexity and technical challenges of the underlying technology. The average cost for Analytics and Tools development may range from $25 to $50 million dollars, but the range can go from as little as $5 million for simple follow-on products, up to $100 million for complex genetic-based technology requiring large human clinical testing. However, Treatments can cost upwards of $800 million to $1.2 billion to move from research and development, clinical trials, and finally to the market.

Product Development Stages

The product development path is divided into *Product Development Stages.* The use of stages is a convenient way to manage and monitor your product development progress. Within these stages, there are value-enhancing steps called *Product Development Milestones,* (discussed later in this chapter), which are as numerous and varied as the products themselves. Product Development Milestones are value-enhancing, and demonstrate progress within each stage. One way to segment product development stages for Treatments and Analytics and Tools are:

Treatments	Analytics and Tools
Basic and translational research	Basic and translational research
Development (in-vitro testing)	Proof of concept
Lead/process optimization	Prototype development
Pre-clinical testing (animals)	Clinical validation
Clinical testing (phase I, II, III)	Regulatory review
Regulatory review	Scale-up and manufacturing
Scale-up and manufacturing	Marketing and growth
Marketing and growth	

Development Stages for "Treatments"

Treatments: Basic and Translational Research Stage

During the Basic and Translational Research Stage, scientists take basic research and begin the translation into a product that is valuable to a target market segment. Researchers take already existing scientific evidence supporting the product idea and advance the science further to confirm there is tangible product potential. Examples may include work on a small molecule therapeutic, a cloned human protein, or a monoclonal antibody with activity that supports the possibility it could work as a treatment in humans. During this stage the resulting work must demonstrate that the treatment clearly and reproducibly causes a measured effect in the target assay and that it represents a good surrogate for a human biological system.

For instance, if the plan is to develop a treatment for Alzheimer's disease, the company may want to demonstrate that the treatment prevents the formation of amyloid plaques in an in-vitro assay, interferes with a metabolic pathway leading to formation of these plaques in the brain, or elicits an immune response to the beta amyloid peptide. There are many ways to demonstrate plausibility for a potential treatment, but the stronger the evidence, the more enthusiasm others, including investors, will have for this project. The key is to demonstrate that the product idea has scientific merit, and that the research produces sufficient evidence to move to the next stage where more money and time will be spent.

As universities and research institutions seek ways to increase their technology value, many are beginning to perform translational research in their own research laboratories. By doing this they reduce technology risk and provide a better license opportunity at advanced development stages where the university can derive more for the license. Wherever the translational research is performed, the organization must demonstrate solid evidence this research can lead to a viable product.

Any research that leads to a significant product opportunity is certain to have others that will also working on a similar opportunity. As a side note, an important but many times overlooked practice is good documentation during early product development research. When developing a product, make sure the scientists and technicians keep good laboratory notebooks and document their experiments and discoveries. Be sure notebooks are routinely cosigned and also be sure that to have an archiving system or use an electronic notebook system. During product development, the company will be filing additional patents on new processes and discoveries, and these laboratory notebooks become vital documentation. Notebooks also serve as legal documentation for patent prosecution, interference, and litigation. From my early work, the laboratory notebooks I kept were used as evidence in a patent interference proceeding related to experiments performed many years prior while mapping a neutralizing antibody epitope to HIV. These notebook experiments were admitted as evidence supporting patent prosecution and an interference proceeding. Good laboratory documentation practices will help protect the company's research investment and inventorship priority.

Treatments: Development Stage

The Development Stage for Treatments can be quite lengthy. Although the time spent here decreases in proportion to how well the disease is understood medically, and how well the science is developed that forms the product idea. Prolonged development stages occur when there is scientific uncertainty surrounding a new disease, or when the disease has an unclear etiology. In 1993, when AIDS was first described in medical journals, it was unclear how the HIV virus precipitated this condition; consequently the Development Stage for AIDS vaccines and therapeutics was extremely long. HIV vaccine development was further complicated by the uncertainty of how the virus escaped detection by the human immune system. Even today, none of the early vaccine development programs have produced efficacious products because they were based on the presumption that traditional vaccine strategies would work with HIV. As researchers elucidate underlying causes of complex diseases and therapeutic mechanisms-of-actions are better understood, development stages for products become shorter. When considering a biotech product development pathway, assess the level of scientific uncertainty for the disease condition, and inventory whether the causative agent or mechanism of the disease is well characterized. Knowing this will help gauge a realistic development timeframe and the impact on funding and financing needs.

The Development Stage goal for most products here is to leverage the translational research and produce a compound or molecule, or family of compounds or molecules, that can reasonably be hypothesized to treat the targeted disease or condition. On the basis of surrogate assays that are related to the disease or condition, the research must produce the strongest possible evidence that this product will be valid in humans. To complete the Development Stage, one needs multiple studies demonstrating in-vitro (in cells), that the use of some version of the product, preferably in its most purified form, consistently produces the response sought, and that it is significantly improved over any treatment currently in existence. Companies can spend between two and five years or more at this stage depending on the level of understanding of the disease being targeted, and the understanding of the mechanism involved in triggering this disease. The goal here is to produce strong and convincing evidence that the product has scientific merit to treat the condition or disease, and the scientific evidence must be compelling enough for investors to continue funding to advance to the next product development stage.

Treatments: Lead/Process Optimization Stage

During this stage the therapeutic will be optimized and/or a reproducible pilot process for a biologic will be developed to produce or express the product. For a small molecule therapeutic, numerous variations or modifications of the molecule will be examined for improvement in efficacy or reduction in toxicity. For a biologic, the focus will be on developing a production or expression process that reproducibly produces the same molecule yielding the same testing results. Process development may involve examining different expression systems for posttranslational modification differences such that the molecule functions identically as the human version. For instance, tissue plasminogen activator (r-tPA) was examined in multiple expressions systems from bacteria, insect cells, and mammalian cell lines to find the one that produced the best glycosylation patterns supporting activity and stability of the molecule, until expression in CHO cells were eventually chosen.

It may go without saying, but be aware of product impurity issues. Minor impurities, byproducts, or even stereoisomers may trigger alternative mechanisms-of-action or produce uncertain results when tested under minimally defined testing systems, and these may even be inhibitors or effectors of alternate reactions in a biological process.

Byproducts can confound or produce varied results. Be sure to establish purity and characterization criteria for a product at various testing stages when determining efficacy and toxicity. Sometimes a particular assay system itself may mask an effect that would be seen in human testing such as an endotoxin response. It is important to mimic as closely as possible the biological system in which your product is intended to be used.

If a molecule or biologic can be produced or expressed with differing properties, be aware that sometimes one may be selectively produced at the laboratory scale, and the other may be preferentially produced in a scale-up mode. During early stage product development, there may not be the expertise available that large pharmaceutical companies have, so it is important during this stage to seek out guidance from experienced individuals in the industry. Presenting results at scientific meetings provides credibility and is helpful because it comes with feedback from others in the field. First share all data with your patent attorney before disclosing information, as you want to balance disclosure with the need to patent your discoveries.

Making progress during this stage is critical. This is typically where companies either run out of money or discover that the development timeframe will take longer than expected. If key funding milestones are not reached, weigh the value of staying the course to complete the development stage, or modify these plans to create an alternate fundable milestone. Sometimes an organization is caught flatfooted and has no alternatives other than to seek additional money at a substantially decreased value, known as a "down round," where the valuation of your organization is lower than it was after the last round of funding (see Chapters 8 and 9). Careful planning and monitoring of progress can help reduce desperation which often occurs during this stage of product development. The goal at this stage is to produce a lead compound with reproducible efficacy in-vitro, and/or develop a reproducible production process that produces the active biologic or molecule of interest.

Treatments: Preclinical Testing Stage

The preclinical testing stage produces a critical decision point to determine whether or not to proceed into human clinical testing. The evidence for proceeding must be strongly compelling and support potential efficacy in humans. A company will be testing its drug or biologic for safety and efficacy in cell-based assays (in-vitro) and in selected animal models (in-vivo) that mimic the disease condition they are targeting. The company will also be conducting extensive testing to demonstrate safety before moving into human clinical studies. The types of testing include various chemical and biological testing, pharmacology testing, and toxicity and safety studies. For example, when developing a drug for the treatment of solid tumors, it will be necessary to show in-vitro and in animal studies, consistency in tumor response, and demonstrate some acceptable measure of improvement whether it be survival benefit or other surrogate endpoints.

The studies conducted at this stage require experienced personnel in toxicology, drug metabolism, structure-function relationships, and thorough characterization possibly X-ray crystallography, sequencing, and biological and chemical assays. Most of these activities should be contracted out to keep the company infrastructure at a reasonable level. There may be the temptation to significantly expand the capabilities of an organization because there are funds in the bank and the work needs to be done. Before doing this, determine the core competencies that make up the company's strategic assets. If a particular activity is within the core competency, expand these and carry out the activities in-house. If an activity is not a core competency, do not consider anything more than subcontracting the service. Unless an activity is inexpensive and requires little human capital, and it will *always* be needed, and the company can do it cheaper, faster or better – don't. Some activities during this stage are dynamic and should be reassessed periodically, because what was best early on may not be ideal six months or a year later.

When developing a product for market in the US, beyond this point the FDA will be engaged. To begin clinical testing in humans, the submission of an Initial New Drug application (IND) along with all the supporting preclinical data is required. The preclinical testing requirements to move to human clinical testing will vary depending on the treatment and indications, so it is a good idea to hire a qualified regulatory expert experienced with the same indication, and have them perform a "gap" analysis of all requirements before proceeding. The regulatory expert should be someone who has worked with the FDA, and will be someone who helps write your IND, they can also help guide in the selection of help during the clinical testing stages.

Treatments: Clinical Testing Stage

Taking this step moves the company into one of the unique barriers to entry for products in life science industry. Chapter 13 outlines additional detail in the steps from filing an IND though clinical trials and regulatory approval. Refer to that chapter for additional information on each of these remaining stages.

The Clinical Testing Stage requires an enormous amount of capital and will take many years to complete. By proceeding into this stage the company is committing, in a sense, to a do-or-die situation. A development-stage biotechnology company enters into clinical trials watched by an extremely large and interested audience, whereas pharmaceutical companies proceed through clinical stages without as much fanfare. When a pharmaceutical company's drug fails, everyone is greatly disappointed, but the company has the staying power to move on and develop other therapeutics in their pipeline – whereas a development-stage biotechnology company does not get many second chances. Carefully and cautiously plan this stage of development, and seek all the guidance possible from industry and regulatory experienced individuals.

Managing and executing a clinical testing strategy resulting in an approvable drug is challenging even for experts. There is a need to enlist the best expertise to help manage product development from this stage forward. Contract Research

Organizations (CROs) are valuable partners and they bring various expertises. Here are some ways to utilize CROs to manage the clinical testing and regulatory process:

- Completely contract everything out, as turnkey to a Contract Research Organization that has been carefully selected.
- Hire an internal expert or consultant to oversee this, but still contract out everything as a turnkey to a CRO.
- Hire internal experts or consultants to manage and monitor this process while contracting out subportions of clinical testing to selected CROs.
- Hire all of the expertise and perform all the functions in-house.

However a company handles the clinical testing phases, seek experts that have direct experience in leading a successful clinical program. Make sure they have clinical testing and product approval expertise in the same disease category that the company will be seeking approval, although it is not necessary they have experience with a product having the same mode-of-action. Expertise in the same disease area is essential because the requirements and endpoints the FDA seeks can be vastly different between disease conditions. For instance, in some cancer treatments, an acceptable endpoint may be anything from tumor size regression to direct survival benefit, whereas for Alzheimer's disease it may be surrogate endpoints such as improvement in cognitive function. Surrogate endpoints are substitute outcome measures that are more easily measured or achieved than the real measures, and are of practical importance, and represent a reasonable substitute. Knowing FDA acceptable endpoint goals lets one design and plan studies to collect the necessary information to best support a product approval application.

Most biotechnology companies do not have intentions on completing the entire Clinical Testing Stage alone, but anticipate finding a pharmaceutical partner to license the drug or molecule to either take over the remaining responsibilities after completion of Phase I or Phase II clinical trials, or to jointly share in these responsibilities and costs. Regardless of what the intentions are, be sure that a clinical testing strategy is aimed at producing data that supports the drug approval application and market need. Without a good strategy that results in the clinical testing data necessary for an FDA approval, any hopes for partnering will be short-lived. Although the clinical testing process proceeds in separate phases have a complete study plan that outlines what you expect to accomplish throughout *all* phases of clinical studies *before* filing an IND.

Pre-IND Meeting

When preparing for human clinical testing, the company will request a Pre-IND meeting with the FDA. This meeting gives feedback and provides understanding of what the FDA will be looking for from the clinical studies. This meeting is to share the company's clinical trial plans and get FDA's feedback on appropriateness for achieving drug approval. Realize, however, that following their guidance does not guarantee approval.

Investigational New Drug Application (IND)

To begin human clinical testing, the company must file an Investigational New Drug Application (IND) with the FDA. The information requirements for filing are outlined on the FDA website. Once an IND is filled, the FDA has 30 days to respond to the application, or the company may proceed into clinical testing. If the FDA has questions, they will respond during this 30-day period and place your application on "Clinical Hold." This just means that the company cannot proceed to human testing without first satisfying all the FDA's questions.

Phase I, II, III Clinical Studies

Once the FDA is satisfied with the data package and clinical testing protocol, the company may begin Phase I testing in humans. During Phase I clinical trials, the investigational drug/biologic will be administered to a small group of healthy individuals. In some instances, the investigational drug/biologic may be administered to diseased individuals.

If there are no safety issues raised in Phase I, the company will apply again to the FDA to move into Phase II clinical studies where they will be administering the drug/biologic to a larger group of individuals for whom the drug is intended for treatment. Upon successful completion of Phase II clinical studies, there will be significant interest in your product from pharmaceutical companies that may have been following the company; at this point, the valuation of the company will have significantly increased.

Proceeding to Phase III clinical studies is accomplished in a larger group of patients that are intended for the treatment of the disease. Phase III is a pivotal study, which is usually double-blinded, placebo-controlled and randomized, where it can assess efficacy, dosing, and further examine safety. At the conclusion of these studies, there will be an overwhelming amount of data to be sorted and analyzed.

One universal challenge to completing all human clinical studies is recruiting qualified patients in a timely manner. Competition for patient recruitment is high even with large incentives to participate. If an appropriate number of patients are not enrolled, the therapeutic or biologic cannot be properly evaluated, and progress toward regulatory approval is halted. Because of the patient recruitment issue many US biotechnology companies are conducting clinical studies in other industrialized countries that have greater access to more patients for their particular treatment. However, all industrialized countries are beginning to experience these same recruitment challenges. There has been an increasing shift toward clinical trials in developing countries because they provide a new supply of patients, better patient compliance, and they have physicians who have been trained abroad. If this is the company strategy, be sure to consult the FDA about the type and amount of clinical data they will accept for support of an application for approval. When approaching patient recruitment, realistically plan for the recruitment time, recruitment costs and your ability to reach the numbers of patients needed to complete the study. Since it is likely that more than one indication will be targeted, the clinical trial costs will multiply. It would not be unusual to see recruitment and testing costs range between $30,000–$50,000 per patient depending on the indication being studied. A CRO can give more precise estimates of clinical trial costs.

Treatments: Regulatory Review Stage

To receive regulatory approval in the US, the company will submit a New Drug Application (NDA) or a Biologic License Application (BLA) with all the preclinical and clinical data in support of claims and indications-for-use of the drug or biologic. This is the culmination of all development work. Be sure to retain the best regulatory and clinical expertise with direct experience working with that particular FDA division. For products that qualify, there are alternate regulatory filing pathways that provide a faster review time, but remember "faster" is a relative term. There is a Fast-Track, Accelerated Review, and Priority Review designation that that can be requested. See Chapter 13 for more information.

Once an application is filled, there will be a multitude of questions and responses. This review period is lengthy and generates reams of paper trails, though *electronic* filings may soon become standard. This lengthy review process may conclude in a panel meeting convened by the FDA with experts in the field of your treatment. At the conclusion of the panel review they will vote on a recommendation for approval. This recommendation is sent to the FDA, and they then make a decision on the approval of your product. The statistics throughout the entire drug development process is staggering. Fewer than 10 of 5,000,000 compounds tested make it through to animal studies, and only 3 make it into human clinical trials, and only 1 ends up being approved.

Treatments: Scale-Up and Manufacturing Stage

Reaching this stage means the company has been successful in developing a product shown to be safe and efficacious, and received regulatory approval for marketing. There is an entire process development science for scale-up and manufacturing, which we will not go into here. Suffice it to say that processes and molecules produced in the lab do not always scale-up to manufacturing levels in a straightforward manner. There are different limitations and efficiencies that impact the recovery and production of the same molecule or compound from a lab process and a scale-up process. It is important to be aware of these issues during the Development and Lead Optimization stage. For instance, one may be able to easily purify a molecule or compound for testing by HPLC in the lab, but this does not make an ideal manufacturing process. Keep one eye on the downstream scale-up issues, as this will help economics and manufacturing when reaching this stage. Oftentimes, small biotech companies may contract the scale-up process and manufacturing to specialty bioprocess or manufacturing houses. It is still important to be aware of the potential issues with scale-up and manufacturing of a product as it will impact progress.

Treatments: Marketing and Growth Stage

Growth can also bring new problems that may not have been anticipated. Uncontrolled growth is challenging to any organization, and can result in problems that are not easily fixed such as when the company was small. As the company grows changes must go though several layers of approval and several departments for collateral effect.

The marketing strategy is now a reality. The future success of the organization depends on reaching profitability as quickly as possible. Biotech companies that reach this stage enjoy a measure of celebrity status as long as there continues to be a growing market for the product and they demonstrate increasing revenue. Experienced biotech companies have occasionally run afoul by going beyond FDA approved marketing claims by unlawfully promoting off-label usages for their products. Though it is legal and ethical for physicians to prescribe drugs for off-label usage, it is unlawful for company representatives to promote these usages in their marketing efforts. Most all existing drugs find valuable alternative uses, but it is unlawful for company representatives to promote these. Many times these off-label usages can result in new clinical trials for these additional indications.

Development Stages for "Analytics and Tools"

Analytics and Tools: Basic and Translational Research Stage

Just as for Treatments, the Basic and Translational Research Stage starts by taking research experiments, and begins translating them into something of value to a target segment of a market. A key difference for Analytics and Tools is the expectation to complete this stage and subsequent ones much quicker than for Treatments. During this stage, expect to advance the science as far as possible to demonstrate product development potential. To advance to the next stage, one must prove that the product idea has merit, and the product concept can readily be translated into a product. There are many ways to demonstrate plausibility, but the stronger the evidence, the more enthusiasm others will have for the project.

Creativity is important at this stage. Product ideas originate and develop when researchers think about how to creatively solve a particular problem. Inspiration can occur anywhere, such as the example of when Karry Mullis conceived the polymerase chain reaction while just driving along a California coast highway one evening.

Analytics and Tools: Proof-of-Concept Stage

The goal here is to reach proof-of-concept for the technology or product. This is not necessarily a prototype but rather proof, that can be independently verified, that the idea or concept would be valid in the real world. For example, proof-of-concept for a marker intended to diagnose an early stage disease condition needs to demonstrate that a high percentage of diseased individuals possess this marker, and it is absent or minimally present in nondiseased individuals. During this stage, the assay system used to detect a molecular marker is not expected to be commercial ready. It may still be labor intensive and impractical. However, it needs to be demonstrated that it can consistently detect the marker, even in the presence of potentially interfering substances usually found in these types of specimens.

For instruments and tools, the objective is similar. You want to demonstrate that a version of the instrument can perform the function it seeks to claim. I currently serve as an advisory member for an infrared laser company that is exploring detection of various molecules in breath as indicators of disease. The product proof-of-concept was to demonstrate that a specific laser can reliably detect certain molecules that correlate with asthma and its progression. This goal was not to create a hand held laser that can detect and print a diagnosis, but to demonstrate with high correlation that these molecules associate with asthmatic conditions. At this stage, the instrument was not intended for commercialization, and the output was not interpretable without a Ph.D. in physics or engineering. Once they demonstrated proof-of-concept, they received seed funding to support further development of this product.

Set Proof-of Concept milestones that establish credibility and reasonableness and demonstrate the idea has merit and is worth funding. Your goal is to produce evidence that is strong enough and convincing enough to obtain seed funding, so you can progress to the next stage of product development.

Analytics and Tools: Prototype Development

The goal here is to develop a prototype product that can be tested for performance in some manner, prior to human studies. If the product is an implantable device, this means producing a prototype that can be evaluated, usually in animals, to verify the device functions properly under a variety of conditions. When developing a medical test, the prototype must be tested on the intended use population and/or run by the intended end-user. Ideally, this means completing development of a prototype that is near final or very close to it. Notwithstanding regulatory issues, there could be reasons why one would want to use an early stage prototype and introduce final changes later, if a new technology such as a better chip is available soon. However, be wary of unproven components no matter how wonderfully touted. You do not want to place the future of your product into the hands of others by becoming codependent on their quality and performance. For medical tests, realize that downstream changes may introduce variation or unforeseen chemical reactions that may alter results in an unanticipated manner.

Some prototype development activities should be contracted out to help keep your infrastructure at a reasonable level. As discussed previously there may be the temptation to significantly expand the capabilities of your organization because you have the funds in the bank and you need the work done. Before doing this you should determine which core competencies are your strategic assets. If an activity is not a core competency you should *not* consider anything other than subcontracting the service, unless it is reasonably inexpensive, requires little human capital, and you will always need it, and you can do it cheaper, faster or better in-house – otherwise forget about it. However, if a new function strengthens your core competency, then add to it. For instance, if you are a gene discovery company and your core competency is proprietary methods of discovering genes that correlate with disease, anything tangential to this objective builds on your strength. Whereas, you would outsource, say specimen acquisition or regulatory functions early on in the development of an organization. Do not forget that your functions during this stage are dynamic and should be reassessed periodically. What is best now may not be ideal six months or a year later.

Prototype development time and costs differ significantly, depending on the development objectives. This is the stage when companies typically run out of money, or find that the development timeframe takes longer than expected. The ultimate goal at this stage is to produce a prototype that demonstrates its intended capabilities, and allows the company to move to the subsequent stage where a final assessment of efficacy can be determined.

Analytics and Tools: Clinical Validation Stage

During this stage, the company must have a final product that will be used for commercialization. This is because they must test the product in, or on, the intended use population. For medical testing and devices not considered to be "High Risk," some studies may be conducted on clinical specimens without first receiving an Investigational Device Exemption (IDE) from the FDA prior to testing (companies developing High Risk medical devices must first file and receive IDE approval prior to testing in humans). The goal during this stage is to validate results in multiple populations. Ideally, you will want to have independent validation from other nonassociated institutions conducting the testing. The results from these studies will be used to submit a 510(k) or a PreMarket Approval (PMA) application.

Analytics and Tools: Regulatory Review Stage

During this stage for US marketing, the company will submit either a 510(k) or a PMA application to the FDA. The statutory review time for a 510(k) is 90 days, whereas the PMA review time is 180 days. However, this review time is based on the FDA clock and does not include response time. In addition, the clock can be reset if there is significant reason to support this. During review, expect several rounds of correspondence and questions based on the complexity of the submission and the uniqueness of the product and technology. More information on the regulatory process for medical devices and IVDs is reviewed in Chapter 13.

Analytics and Tools: Scale-Up and Manufacturing Stage

Congratulations! By reaching this stage the company has been successful in developing a product and has received regulatory approval. For commercialization in the US, the manufacturing processes needed to meet FDA regulatory guidance includes Good Manufacturing Procedures (GMP) and optimally ISO certification. There are many excellent resources available that provide a good understanding of GMP and ISO requirements. I would recommend that entrepreneurs familiarize themselves with these before reaching this stage.

Analytics and Tools: Marketing and Growth Stage

The marketing strategy has now come into reality. The future success of the organization and its ability to become profitable, and ultimately have an exit strategy for investors and a reward for management and employees is at stake. If the company's product is in a desirable market, there will be great potential for marketing partnerships, licensing, or even an acquisition of the company.

Growth also brings new challanges that may not have been anticipated. Uncontrolled growth can be detrimental to a young organization, and result in problems which are not as easily fixed when the company was small. In a small company, changes can be immediately implemented throughout the organization before the end of the day. Whereas now, changes must filter though layers of approval and departments for collateral effect, such that many decisions may not be implemented for weeks or even months. Successful companies will learn to adjust to these changes, because they have reached their goal – commercialization.

Product Development Milestones

Product development milestones are measurable accomplishments that incrementally move a product closer to commercialization and incrementally reduce development risk. Specific product development milestones vary according to the product category. It is important to identify appropriate product development milestones at the outset. When milestones are met, they instill trust between the company and its investors. This credibility is vital should the company encounter unanticipated problems that require additional capital in the future.

For a *Treatment*, examples of value-enhancing, risk-reducing milestones include the following:

- Licensing the core technology from the owing intuition
- Licensing additional patents that provide freedom-to-operate in that field
- Securing new issued patents with required claims
- Establishing a top-notch Scientific Advisory Board
- Completing proof-of-concept experiments successfully
- Determining mechanism-of-action for the compound or biologic
- Improving the compound's efficacy through structure-function relationships
- Successfully selecting a lead compound
- Successfully completing animal studies without any safety, efficacy, dosing, or delivery concerns
- Publishing results in top-tier peer-reviewed journals
- Winning Phase I and Phase II SBIR grants
- Entering into a service agreement with a well-respected company or organization
- Forming a partnership or strategic alliance with a credible marketing partner
- Successful filing of an Investigational New Drug (IND) application with the FDA
- Licensing the product to a pharmaceutical company
- Completing Phase I Clinical Studies with successful results
- Completing Phase II Clinical Studies with successful results
- Completing Phase III Clinical Studies with successful results
- Filing an New Drug Application (NDA) or Biologic License Application (BLA) with the FDA
- Obtaining FDA approval

For *Analytics and Tools*, examples of value-enhancing, risk-reducing milestones include the following:

- Licensing the core technology from the owning institution
- Licensing additional patents that provide freedom-to-operate in that field
- Securing new issued patents with required claims
- Establishing a top-notch Scientific Advisory Board
- Completing proof-of-concept experiments successfully
- Demonstrating feasibility and prototype production of the product
- Completing pilot clinical testing of the prototype or final product
- Publishing results in top-tier peer-reviewed journals
- Winning Phase I and Phase II SBIR grants

- Entering into a service agreement with another entity or organization
- Forming a partnership or strategic alliance with a market partner
- Completing clinical validation studies demonstrating efficacy or improved sensitivity and specificity over a gold standard
- External clinical studies that validate in-house clinical data
- Filing a 510(k) or a Pre-Market Approval (PMA) with the FDA
- Obtaining FDA marketing clearance or a PMA

Choose development milestones specific to your particular product, then work to meet these milestones in a timely manner and within budget. By consistently meeting product development milestones the company will ensure the best opportunity for success.

Commit Your Plan to Writing

A product development plan must be reduced to writing. List product development stages, key milestones, and the results you expect to achieve at each stage. Identify any uncertainty in these plans and utilize project planning software to create a Gantt or PERT Chart of the timeline (see chapter 17 for additional information on project mangement). The plan should show timeframes for parallel and critical path activities along with any slack time. Once a written plan is developed, it can later be modified and updated with any new information. Do not worry about being too detailed at this time; just be sure to capture major milestones and key requirements, along with any specific needs to achieve these. Project planning should be continually updated and modified on an ongoing basis.

Next, estimate the costs to complete each stage of the plan. One helpful way to estimate reasonable costs is to talk with a Contract Research Organizations (CRO), which are companies that assist at any and all development stages of a product. Good CROs will spend the time to talk with a company about their product development questions. The company will most likely need their help sometime during product development or clinical testing phases. Also, look for experienced biotech executives that have worked through each of these development stages and get their advice. Another way to benchmark estimates for total development costs is to access databases such as Dow Jones VentureSource. These databases list privately held companies and their capital raises for each round of financing. From this information, comparable product development costs can be gleaned, along with a total development cost estimate and the number of rounds of capital needed to reach specific product development stages. This information may not be broken down into development stages but one can glean overall costs for development to a particular stage.

Product Development Assistance

A biotech company can never have too much assistance. Explore alternate ways to stretch development dollars by applying for free services available from the federal government. The National Institutes of Health (NIH) has a wealth of resources and many of these programs are free, even to for-profit companies. Consider these programs in addition to grants when seeking ways to stretch development dollars. At a previous start-up company, we utilized an NIH program to screen candidate compounds for antiviral activity using their in-vitro and in-vivo assays. A substantial amount of time and resources would have been required for us to set up these testing models and protocols. With their help we got critical proof-of-concept data, which was used to help raise investor interest. In another instance, we took advantage of the NIH's cancer screening program to screen a group of potential antitumor compounds. The NIH also has assistance in other areas of drug development such as the synthesis in bulk of small molecules, development of analytical methods, isolation and purification, pharmacokinetic/ADME studies including bioanalytical method development, development of suitable formulations, and manufacture of phase I drug supplies, to name a few.

Assistance for preclincial therapeutic and biologic drug development is available from the NIH through a program called Rapid Access to Interventional Development (NIH-RAID). A similar program, Rapid Access to NCI Discovery Resources (RAND) is also available; partnerships are selected on a competitive application basis. These programs are for academic and not-for-profit investigator – initiated studies only, but depending upon the stage of license from your research institution, there may be ways to work together to utilize these programs. The study must be investigator-initiated, but small businesses (less than 500 employees) may still qualify even after they have licensed the intellectual property from the research institution.

These programs are available for both US and non-US institutions meeting certain requirements. Additional information and requirements can be obtained through their website.[1] Certain requirements must be met for each of these services, but if a company qualifies, these services are available to assist in the development of certain drugs that are deemed important to the welfare of the US public. Check out these and other resource opportunities. It will be time well spent, and can accelerate product development, and save precious research and development dollars.

Significance of Winning Competitive Grants

Winning competitive grants provide additional money for product development, but also indirectly gives validation to your technology or product concept. Securing a sizable phase II SBIR grants can significantly boost a company's scientific credibility, not to mention this money is nondilutive (money that does not require giving up equity). If the company does not have a person who is a good grant writer, or those in the organization have not written competitive federal grants before, there are resources available to help. Just remember SBIR grants are not handouts to pay for clinical studies or purchase a piece of equipment. Awarded grants seek to solve scientific problems of broad interest, that could result in a product development.

---

[1]NIH-RAID website: www.nihroadmap.nih.gov/raid/ RAND website: www.dtp.nci.nih.gov/

SBIR grants must be hypothesis-driven, meaning applicants must be asking an interesting research question and be testing a hypothesis. Remember, these grants are ranked and evaluated, most often by academic researchers. Make your research approach intriguing, and demonstrate its significance such that scientific reviewers can see its value. Do not forget that SBIR grant are awarded for projects that could result in an economically viable product, so be sure to carefully outline the commercial value of the research.

Commercialization Help from SBIR

By winning nondilutive funding from a Phase II SBIR grant, a company then becomes eligible for help from the NIH's Commercialization Assistance Program (CAP). This program was started in 2004 and is currently run by the Larta Institute which provides help in finding partnerships, raising money, and assisting in the commercial development of a product. This assistance supports all segments of the biotechnology industry such as specialty therapeutics, diagnostics, medical devices, and other healthcare support services. This valuable resource is a 10-month program available to selected SBIR Phase II awardees. This service provides workshops, training, and mentoring to support an organization with the information and tools needed to improve one's success in raising capital and in progressing through product development. The CAP program is funded through the NIH and its continuance will likely depend on success measures and future appropriations from the federal government. The NIH states their program objectives are to help the biotech company with:

- Developing or improving strategic business planning
- Developing or improving market identification, definition, and research capabilities
- Developing or improving marketing materials, e.g., investor pitch, company brochure, etc.
- Developing a roadmap for, and establishing licensing opportunities
- Developing a roadmap for, and establishing strategic and/or investment partnerships
- Seeking regulatory approvals

You can find more information about NIH CAP program at the following websites: http://grants.nih.gov/grants/funding/cap.htm, and at http://www.larta.org/

Market Development and Business and Organizational Development Paths

There are two additional development paths that run parallel to the Product Development Path we just reviewed. These development pathways should not be afterthoughts – these are *Market Development*, and *Business and Organizational Development* paths. It is just as critical to make progress along these paths (reviewed in subsequent chapters). Most biotech entrepreneurs are somewhat familiar with the product development pathway, but fewer individuals have a good understanding of the market development, and business and organizational development pathways for their company. Market development pathway components include market research, segmentation, targeting and positioning of the market, identifying a value proposition to the target customer, development of pharmacoeconomic models, and a reimbursement strategy, to name a few. The business and organizational development components include establishing the proper corporate structure, adopting good hiring practices, aligning organizational strategy, building the infrastructure, and forming strategic alliances and partnerships. These additional paths must advance simultaneously. Fortunately, they are not as expensive or time-consuming as product development – however, they are just as critical. These two additional pathways must be advanced at appropriate times and be in sync with the product development strategy.

Summary

Vast sums of capital are required to develop biotechnology products. Reaching each milestone and development stage must be completed in a timely manner, in order to support funding events and increase the value of the organization. Achieving these milestones help secure new investor interest, partnerships, and increase the likelihood that subsequent funding events will occur. Careful planning of goals and milestones are critical in order to reach them prior to the exhaustion of existing capital. Think of this situation as driving a busload of people along old roads through the Mojave Desert without an accurate map or a GPS device. Gas stations are neither plentiful nor frequent, so you must be aware of the distance you need to travel, and the amount of fuel you have to reach the next filling station – a reserve tank would be a life-saver. Those that are old enough remember the VW bug automobile having a reserve gas tank lever. This one feature has saved the author hiking trips along the freeway during my teen-age driving years! So, plan carefully, watch your expenses, and work creatively to accomplish your development goals within budget – and plan for a reserve. As you successfully reach subsequent product development stages, you will find corporate partners want to join forces with you to accomplish the remaining development stages essential for commercialization of your product. These relationships can include alliances, partnerships, or collaborations. Partners can also be academic and government institutions, as well as commercial organizations such as pharmaceutical companies or larger biotechnology companies (discussed in Chapter 12).

The product development stages described in this chapter provide a typical development pathway, but it is not specific enough for all products. These stages and suggested milestones serve only as a reference to help develop a specific plan tailored to your product. Take plenty of time to develop a sound research and development plan. Outline each stage and include estimated costs to reach each milestone. Doing this will help reduce the number of unplanned obstacles. When biotechnology companies fail, it can usually be traced to one, or both, of the following reasons:

1. Not having a strategic plan, or having a poor strategic plan
2. Not executing well the planned strategy

In other words – don't "wing it." Invest time in a careful planning process and seek help from those who have done this before. Do not forget the old adage "If you fail to plan, you plan to fail." Remember that plans should be dynamic and frequently updated as progress is made and new information is learned. Be sure to make progress in the *Market Development Stages*, and *Business and Operational Development Stages*. Finally, seek assistance from federal agencies, and do not forget to apply for competitive SBIR grants to support portions of your product development. These nondilutive funds help stretch development dollars by providing alternative funding to reach company goals. Even with the best of plans, an entrepreneur must still be creative, adaptive, and resourceful as they seek to discover new ways to overcome product development challenges.

Chapter 7
Developing a Marketing Strategy

Marketing success is the lifeblood of any company. Everyone knows that a marketing strategy impacts the company's ability to generate profits after commercialization; but understand that it also impacts one's ability to raise money from investors. It is not enough to just develop a product – though it may seem like it should be because of the difficulty accomplishing this in the biotech industry. Before beginning to develop a product, the entrepreneur must show that the product will be accepted by a target audience and that their marketing strategy will accomplish this goal. All too often good product ideas do not get funded because of a poor marketing strategy.

Before developing a marketing strategy, there are a few basic marketing concepts with which to be familiar. In the first half of this chapter, we review some basic marketing concepts and market research tools. For the reader with a business and marketing background, you can skip over this review. However, for those without a marketing background, you should learn these concepts, as it will help explain market issues and provide an appreciation for the challenges in developing a good marketing plan. Occasionally, marketing concepts can be difficult to understand – to simplify them, I have included examples of well-known consumer products and household brands to show their application. Later, we apply these marketing concepts to two biotech products, and show how these tools can be used. This exercise will enable one to start developing their own product marketing strategy. There are many good books on marketing, and a few have been listed in the Additional Reading section for those who want to learn more about this under-appreciated and misunderstood aspect of commercialization.

What Is Marketing?

Marketing is both an art and a science. Most people think of marketing as just an art, yet there are sound principles and market research tools that will allow development of an excellent market strategy for a product. Marketing is not the same as "sales," which is the execution of the market strategy – sales are the result of good

C.D. Shimasaki, *The Business of Bioscience: What Goes into Making a Biotechnology Product*, DOI 10.1007/978-1-4419-0064-7_7,
© 2009 American Association of Pharmaceutical Scientists

marketing. My favorite definition of marketing is: "Marketing is meeting needs profitably." To help understand what marketing is *not*, I have elaborated on a few common misconceptions.

Misconception 1: What Is So difficult About Marketing? You Just Make It and Sell It!

Most entrepreneurs think their product is so wonderful that everyone will want it because of the fantastic things it can do – think again! The rapid adoption of the personal digital assistant (PDA) was often associated with Palm Pilot and the successful company Palm, Inc. These handheld devices enjoyed wide-spread market acceptance (now these functions are incorporated into our cell phones). These PDAs kept our calendar, stored our contacts, saved notes, and kept us organized, all at our fingertips. Although Palm, Inc. was widely known for the PDA, most consumers do not realize that Apple Computer invented the first PDA. Apple created the Newton, a PDA-like device, which was manufactured from 1993 to 1998, but eventually discontinued it for lack of market interest. Shortly before the Newton was discontinued, an unknown competitor, Palm, Inc. launched its Palm Pilot, developed by first researching the needs of their customers. Palm, Inc. successfully captured a broad market with a second generation product that was originally introduced by an unsuccessful competitor. What did Palm do differently than Apple? Palm developed their product using the Marketing Concept, which we discuss below.

Misconception 2: The Better Your Technology, The More People Will Want to Buy Your Product

There are junkyards full of great technological products developed by companies that were unable to successfully market them in spite of technologically inferior competition. Take for example the first home video recorder developed by Sony in 1975 using their Betamax format. A year later, JVC introduced a competitive VHS format. JVC quickly understood the need to license their VHS format to other companies to gain broad market acceptance, but Sony did not follow suit. The Betamax and VHS formats were incompatible, and the consumer was left to choose.

Consumer demand grew for video recorders, and companies manufacturing VHS video recorders met the growing consumer demand as more studios produced movies in VHS. By 1987, VHS commanded a 95% share of the video recorder market. Gradually movie studios discontinued supplying Betamax format movies, and Sony ultimately discontinued manufacturing its Betamax video recorder. Although Sony was the first mover in this market and the Betamax format was arguably superior to its later competitor, the video tape format battle was won by VHS. For lack of a superior market strategy, Sony lost the opportunity to be the dominant player in home video recording. Hundreds of millions of dollars were deployed in development and manufacture of the Betamax format, which included many products that supported this format.

Interestingly, another format war recently emerged for the high definition optical disk. In early 2006, Toshiba was the first mover to introduce the HD-DVD format and the first HD-DVD player. Sony followed later with its own Blu-ray format and disc player. Sony quickly secured more commitments from movie studios to offer their titles in blu-ray format than did Toshiba. Although technology buffs argue that Toshiba's HD-DVD was a superior format – guess who won this battle?

Misconception 3: I Do Not Need to be Concerned with Market Strategy Now Because I Have Plenty of Time to Think About Marketing Later

Do not be deceived – this cannot be further from the truth. Remember that lots of money is needed to develop a product, and it comes from OPM (other people's money). You will not attract investors unless you can demonstrate that there is a significant market need for your product and that you have a market strategy that gets your product adopted profitably. If you do not believe this, ask a few venture investors, what are the three critical elements they need to see in a biotech company before they invest. You can be sure that demonstrating a real market need, coupled to a great market strategy, will be high on their top three list.

Many business plans are turned away by investors because the entrepreneur could not adequately prove a real market need for their product; or alternatively, the market need was evident, but the entrepreneur could not show how to profitably capture this market. Potential investors do not have the time to dream up better marketing strategies for your business. If you do not understand your own product's market, investors know you will not make good decisions about other facets of your business either. As the sophistication level of your investors increase, the more your marketing plan will be scrutinized for flaws. It is essential you understand the market issues surrounding your product, and produce a well-planned market strategy to address these. Investors do not get excited about a high technology product in search of a market need.

The Evolution of the Marketing Concept

The marketing concept is the philosophy, that companies must analyze the needs and wants of their customers *before* making product development decisions. This is because knowing a customer's needs helps to develop products that satisfy them better than the competition. Understanding how this concept evolved will help one appreciate the significance of the marketing concept in business. There have been at least three major commercialization concepts that evolved during the development of products and services in the US. These three concepts are briefly reviewed below.

The Production Concept

The Production Concept is the philosophy that companies just need to produce their products at a reasonable price, in sufficient quantities, and the products will sell themselves. During the Industrial Revolution, the Production Concept worked well because at that time there was an acute need for basic necessities at affordable prices. Indeed, the limitation for companies was the ability to produce their products efficiently, and in sufficient quantity to make it affordable for everyone. As cost-effective manufacturing processes became available, products did indeed sell themselves. As manufacturing efficiencies continued to improve, many items previously unattainable by consumers suddenly became affordable to the masses. Henry Ford epitomized the Production Concept with his comment about selling Model T automobiles to his customers "You can have it in any color you want, as long as it is black." Over time, all competitors' production capabilities improved, and soon, cost-effective manufacturing became standard. By the late 1940s and early 1950s most basic needs were being met for the average US consumer.

The Sales Concept

As mass-production capabilities became commonplace, and competition for basic necessity products increased – the Sales Concept was born. This was the philosophy that it did not matter if the product was needed by the customer; the focus was on how you could persuade consumers to buy the product. Companies began to work on "selling" customers their products – whether they needed them or not. The goal was to sell their product more effectively than their competition. This became the era of the "hard-sell" and the door-to-door salesmen. As consumers got wiser, the hard-sell no longer generated the revenue that organizations required, and smart companies searched for other ways to be competitive.

The Marketing Concept

With increasing competition, companies searched for better ways to grow revenue and sustain an advantage over their competitors. Companies began identifying unmet customer needs and then developed products that satisfied them – the Marketing Concept was born. Success came to companies that better understood the needs and wants of their customers, and provided the precise products they desired. The Marketing Concept evolved around a key focus of customer satisfaction. If Henry Ford's comment about his Model T could be modified to capture the Marketing Concept, it might sound something like this: "We can color coordinate your automobile upholstery to the leather trim of your choice and pre-program all the drive settings, temperature and satellite stations to each driver's preference in their key fob. This luxurious driving experience comes in coupe, 4-door and sedan, and a limited edition convertible model, each comes in 10 exterior colors and 5 coordinating upholstery colors, and 15 upgradeable features. Just tell us what you want, and we will have it detailed and delivered to you today by 5 p.m."

The Marketing Concept is about understanding the needs of customers even if they do not express them. A very successful Marketing Concept company is Proctor and Gamble. A large part of P&G's target market segment is women aged 30–45 who purchase household supplies. To understand the needs of their target market better, P&G not only asks these women questions about their household needs, but they also visit consumers in their homes and even compensate customers to shadow them throughout the day. Based on observing and asking their target market segment about their household needs, P&G then develops products that meet these needs better than their competitors. For instance, a product developed from this type of market research was the Swiffer®. This product created a new category that was neither a broom nor a mop, yet accomplished the same functions – clean and capture dirt – but was less time-consuming, easier to use, did not require water, and could be stored in a small space. The Swiffer met the needs of their target market so well that in the first year of introduction, sales of this product were over $200 million dollars. The Marketing Concept is the most effective marketing strategy, yet many organizations neglect to properly define their target market segment, and fail to understand their needs.

Understand and Appreciate the Market for Your Product

Not only is it vital to understand a target market's needs, but also it is imperative that a company communicate their product's value, because a product's benefits are *not* intuitively understood. To accomplish this, the entrepreneur must first appreciate something they may not believe – that investors and potential customers may not be as enthused as they are about their product. Most biotech entrepreneurs, especially those with a scientific background, have a tendency to underestimate the market risk for their product. This comes from viewing marketing too simplistically. There is an erroneous belief that once a product is developed, everyone will want it and it will sell itself. This lack of marketing appreciation may arise from limited experience with actual marketing challenges. Learn the market issues for your product, and work through a reasonable market strategy. Formulate the product's value proposition, and identify a plausible marketing channel to get the product to its target market.

For entrepreneurs developing a therapeutic, the market strategy should unequivocally demonstrate a real market need for the product; however, it is not necessary to implement an actual go-to-market strategy. For those in the diagnostic, medical device, biological reagent, and molecular testing business, they must demonstrate a strong market exists for their product *and* they must implement an actual go-to-market strategy because they will be doing this sooner than their therapeutic colleagues.

Marketing Tools

In the following sections, we briefly review tools used in developing a marketing strategy. All marketing can really be summed up in Segmentation, Targeting, and Positioning. Each of these concepts is interrelated. There are many other marketing principles, but it could be argued that they are just refinements of these principles. For our purposes these three will suffice.

Segmentation

Segmentation is the dividing of a population into smaller homogeneous groups based on their needs, wants, desires, and purchasing habits – this is your target market. Segmentation also identifies a group of individuals that can be reasonably reached simultaneously. Potential users of a product must be segmented to identify those with the highest motivation to purchase or use a product. Without segmentation, it is impossible to target them appropriately. The more homogeneous the needs, wants, desires, and purchasing habits, the more likely a segment will respond in a similar manner. Segmenting a population is accomplished by using at least three filters: Geography, Demographics, and Behavior/Usage. Each of these filters may not be applicable to every product, and may have differing importance for segmentation. As products and their uses become more sophisticated, including medicines, diagnostics, and medical devices, more thought is required to properly identify and define a target market segment.

By Geographic Location

Segmentation by geography may or may not be reasonable for a biotechnology product, but it is important to determine this at the outset. Segmentation by geography can be illustrated in the market strategy of See's Candies. In 1921, See's Candies began in Los Angeles, California and has been a favorite of West Coast customers for decades. Their manufacturing plant is located in South San Francisco, and they deliver a quality and freshness that became a significant component of their brand value. See's quickly learned that long-term storage and lengthy shipping cycles took a toll on flavor and taste. If they shipped and inventoried their product across the U.S. for year-round consumption, the product quality declined, and they knew they would lose loyal customers. Therefore, they limited their marketing to neighboring states around their West Coast operations. Geography limited See's' early commercialization strategy. Later, See's sought to overcome their geography limitation, and expand their business to other market segments. See's developed a geographical marketing strategy for the Christmas season, and shipped only prepackaged candies to mall kiosks where they knew they could sell inventory quickly, and preserve quality and flavor without damaging their brand. This marketing strategy became so successful that they expanded this strategy to other states, where former West Coast residents were transplanted having fond memories of See's Candies. Although See's had geographic limitations, they overcame them with a market strategy that did not require setting up and maintaining costly manufacturing facilities across the country.

For biotechnology products, geography may not pose such a problem although there may be some exceptions, such as with a frozen formulation for a new nasal spray vaccine, where distributors must have freezer storage and transport capabilities. Geography can be a marketing consideration in world-wide markets and between countries, based on locations for manufacture, testing, or usage of labile products or services. However, these issues usually can be overcome with a good market strategy.

By Demography

Demographic characteristics are easy to identify, and include qualities such as age, sex, family status, education level, income, occupation, and race. For products that are targeted toward industrial markets, some of these demographic qualities are also applicable, such as age of the organization, size of the organization, and type of industry organization. Biotech products that target physician group practices can include demographic qualities such as, type of practice, specialty, number of physicians in a practice, number of patients seen per day/week/month/year, age of the practice, and age of the physicians.

By Behavior and Usage

This can be the most difficult characteristic to understand, but often it is the most effective and successful method of market segmentation. Segmentation by demographics alone may be too broad of a segmentation pattern since all customers of a particular age and gender – say men 40–50 years of age – do not drive the same type of car, do not purchase the same type of clothes, and do not always want the same type of medical services. Whereas segmentation by behavior coupled to demographics, targets similar needs and wants, such as the segment of women 20- to 30-years old who love cats and shop at Wal-Mart, or men who consume Mexican food and attend an East Coast University. Other segmentation characteristics include consumer buying patterns, a loyalty to a particular brand, usage patterns, lifestyles, and attitudes about certain classes of products. Physicians can also be segmented by behavior characteristics such as drug prescribing practices, usage of practice protocols for a particular disease, and where they send their referrals. Patients can be segmented by preference of a product, usage of a type of medication, and amount of expenditures in a medical category. However, the more detailed the segmentation, the smaller the target market, so realize there are opposing forces that either expand or further refine the segment.

Targeting

Targeting is the narrowing down of a profile for the ideal customer. It is the process of selecting the best market segment, and it includes how to approach them, and the method in which they are approached. Targeting lets one focus and tailor their product to the specific preferences of their best customers. Targeting leads to a product's positioning and it formulates part of a marketing strategy.

Positioning

Positioning is the way one wants their potential customers to view their product. Positioning conveys a product's value to the buyer, and it is the way the product solves the problems of their customer. The more effective a product's positioning, the easier it is for customers to understand the product's value. Positioning has a lot to do with a customer's perception of a product – you are positioning an image of true value associated with your product. Positioning requires that one differentiate their product from other products in the market. In order for potential customers to recognize your product, you have to position it in a way customers can differentiate it from all others. All products must differentiate. It is impossible for a product to be all things to all people, and still have customers value the product. For instance, some people enjoy hot tea, and others like iced tea, but one cannot capture both markets by giving everyone a single product – lukewarm tea. With biotech products, positioning utilizes benefits, value, and solutions to a patient's problems, rather than just attributes of status, image, and experiential benefit, such as possible with consumer products. Biotechnology products must have an underlying improvement in effectiveness and performance in order to be successful.

Branding

Branding is the personal sense of benefit, value, and desirability customers associate with having a product. Companies sell products, but they are really marketing brands. Branding is more than just a product trade name and its recognizable product packaging – it is all the information evoked in the mind of the consumer when they encounter a product.

Companies spend many years carefully creating brands that evoke value in the minds of their customers. For instance, when someone thinks of the safest automobile, they may think of Volvo. When one thinks of automobile reliability, they may think Toyota or Honda. When one thinks of luxury, they may think Mercedes-Benz or Lexus. Notice that Toyota's brand is not associated with luxury because their target market customers do not want to pay $60,000–$80,000 for a Toyota automobile. Toyota knew that they could not get their target market to pay high prices for a "luxury Toyota," so they successfully created an upscale Lexus brand using the same or similar parts. Brands are created over time by consistently delivering products that are in line with a brand's promise.

For biotechnology products, branding is strongly associated with the tangible medical benefit a product provides to the patient, or physician, or medical researcher. There are other factors that help create the brand's value, and these can be enhanced if there is first a tangible medical benefit to the product.

Market Research and Assessment Tools

Market Research is the foundation upon which to build a market strategy – it is obtaining information about a product's potential customers. Market research includes all methods of gaining information and insight about the market for a product, and the needs of their customers. For scientists, engineers, or physicians, think of this process as a graduate level thesis, proposing a hypothesis and examining data pertaining to the market needs of the customers in question. From this information, one will formulate conclusions in a stepwise manner just as one would conduct scientific experiments – building upon proven theories before arriving at conclusions. Market research may be gathered from medical providers, specialty physicians, insurance companies, researchers, or patients with a particular disease or condition. Be sure to conduct thorough market research, gathering as much data as possible.

When studying and interpreting market research results, be sure to guard against selectively seeing and hearing only things that confirm a bias for market acceptance of a product. Be your own naysayer, and think of reasons why a target market will not want this product, and why it may not receive medical reimbursement. By doing this, one can then formulate many answers to investor market questions before they are even asked.

Primary and Secondary Market Research

Market research is classified into "primary" and "secondary" market research. Primary market research is a first-hand account of information, such as talking directly with customers and getting answers from them, conducting focus groups, or commissioning a study by a market research group. Secondary Market Research is information obtained by and from others, such as published studies, or commercially available competitive market information. Because all market research has some bias, knowing a bit about polling and population statistics can be helpful.

Market Size and Opportunity

All companies and their investors want to target the largest possible market for their product or service. Venture capital likes to see large market potentials for a biotech product, usually in the range of a billion dollars. Market potential or a market opportunity is considered to be all *potential* customers for a product.

Sales Projections

It is not enough to just state your market potential to investors; the entrepreneur needs to estimate how much of the potential market will be captured with their market strategy during a period of at least five years. This projection is referred to as a *Sales Forecast*, *Pro forma*, or *Management's Projections*. Sales projections are used to show the projected impact of your market strategy, and the attractiveness of your market. There are two ways to develop a forecast: one is the *top-down* approach, and the other is a *bottom-up* approach. The top-down approach is a first approximation, but it is not built by real drivers of demand. The top-down approach is done by first taking the market potential, and estimating a market penetration percentage based on various assumptions, including penetration rates by other companies in adjacent markets. The bottom-up approach is a methodical approach that takes key assumptions supported by data, and builds upon these and other tangible factors to calculate a resulting revenue stream.

For instance, for a company developing a diagnostic test for a particular condition that will be ordered by an OB/GYN for one would start by showing the number of OB/GYNs in the country, then research the number of patients with this condition that any one OB/GYN sees in a day, a week, or a month. Then one makes reasonable assumptions on what percentage of patients with this condition the test would be ordered based on this market research. This way, an investor sees what assumptions are made, and they can agree or disagree with them. However, they cannot disagree with the number of OB/GYNs in the country, or the average number of patients with this condition. It is worthwhile to produce sales projections by both the bottom-up and the top-down methods to arrive at the best estimate of revenue generation.

Be forewarned, sales projections are always suspect by investors. They have seen too many instances where entrepreneurs make up wildly unrealistic projections. Most investors rightly discount all sales projections in a business plan. There is no way to overcome this mental exercise by potential investors. The best thing to do is to provide realistic projections, and list the key assumptions used to arrive at these projections. Show investors that the projections are supported with solid data and logic. Occasionally, there is no data available to support an assumption. In this case, draw parallels from another market or product, but be sure to acknowledge these limitations, and provide reasons the estimates are logical. Do not make unrealistic assumptions or provide unsupported estimates – poor assumptions undermine your credibility in the eyes of investors.

When preparing a pro forma, it is wise to prepare a worst case scenario – one in which some of the key assumptions are not met. For biotech products other than therapeutics and biologics, you will be in the market sooner. If you are counting on precisely hitting your sales projections and your cash is planned to the exact month, slightly missing your projections means that you will run out of money. Be sure to build pro forma projections using various scenarios, and you will end up with better estimates for your sales projections.

Biotechnology Products Typically Have Three Customers

Marketing to a single segment of customers is hard enough, but to make matters more difficult, biotechnology products must gain acceptance from multiple customers to receive broad acceptance. For biotech products, the Patient, the Physician and the Payor, all play a key role in the gaining of market acceptance and success. The ultimate end-user of most biotechnology products is the patient, but usually the patient has less influence in the product selection, and the physician usually makes or shares in the final decision. In the US, the payor must also find enough value in the product to provide reimbursement. Without reimbursement for these products, there is no realistic method for wide-spread adoption of most biotechnology products in the US.

Who Is the Patient?

The patient for a biotechnology product is usually easy to define. If your product is a drug-eluting stent, your patient-customer is an individual with coronary artery disease wanting to avoid bypass surgery. If your product is a new statin drug then your patient-customer is someone diagnosed with high cholesterol that fits the indications-for-use of this drug. If your product is an infectious disease diagnostic, your patient-customer is a person presenting to a clinic, hospital, or doctors' office, with the disease symptoms.

Who Is the Physician?

Determining the physician can be a bit more complex than identifying the patient, as multiple specialty physicians and medical opinion leaders are involved, and each may have a differing interest in your product or service. For the drug-eluting stent, the physician-customer is an interventional cardiologist, who will be inserting the stent into the appropriate patient-customer. For the statin drug, the physician-customer may be a general practitioner, an internist, a cardiologist, a geriatrician, or even a physician's assistant. For the infectious disease diagnostic, your physician-customer may be any physician, but generally there will be a core group of them that will see more cases of this particular disease than others. For instance, if your test is for Respiratory Syncytial Virus, this disease primarily occurs in children less than 2-years old; therefore, the target physician-customer would most likely be a pediatrician or hospital ER physician.

Who Is the Payor?

There are multiple payor sources, such as private insurance (large, small, regional, national), HMOs, Medicare, Medicaid, and each have certain requirements to reach a reimbursement coverage decision. In the US, we tend to have a sense of entitlement for medical care and usually do not expect to be paying much out-of-pocket for healthcare services. This means that to get most products or services adopted in the medical field, you will need to secure medical reimbursement. To get payor coverage, your product must provide a significant benefit to the payor and to their patient. Many private insurances rely upon medical reimbursement evaluation groups for assessing reimbursement decisions – targeting these groups may save you time and money without having to deal with each payor individually. You will need to spend sufficient time examining medical reimbursement strategy issues to get wide-spread adoption of your product, as this is a critical component of success for biotechnology products. Do not presume that these issues are ones you deal with *after* your product is developed. Smart investors will not fund projects that have no reasonable chance of getting medical reimbursement coverage, so be sure to have a good understanding of this as you raise capital.

Multiple customers make biotech product market research and development more complex than in industries where there is only one customer. You must carefully consider each of these customers because you cannot market your product successfully without acceptance from all three.

Five-Step Process to Applying Market Concepts

I have listed below, five steps to help you begin formulating your marketing strategy. There are many more aspects, but those outlined will help you get started developing your own marketing plan. To reiterate, you must first have already determined that there is an acute need for your product, which is not served by other alternatives or substitute products. This is even more important than the market size. Niche marketing categories are seen as opportunities to capture a ready market for a product that would be expected to grow in demand and usage because of the acute need for such a product.

Step 1: Determine Your Potential Market Size

First determine the potential market for your product. This is the largest group of customers that could conceivably purchase or use your product. The total market size represents your *market potential* and is not a sales projection – rarely do product sales get close to capturing the entire market due to many limitations, including competition. By taking the total customer market, you will estimate the total annual market potential in dollars, using your product pricing and other factors such as frequency of use. Investors want to know that your market is big enough to support their investment, and provide the type of return they expect. Most venture capital investors want to see market potentials that exceed $1 billion dollars annually. For instance, it would be foolish for a venture capital investor to invest $30 million dollars into a company if the total market size for the product was only $5 million dollars annually.

There are many ways to determine your potential market size. One way is to evaluate your competitors' potential market, if you have competition. Another way is to evaluate the potential market for substitute products. Remember, you are not using the actual sales of competitors or alternative products; you are estimating the potential market that could be reached with an *ideal* product – which usually is many times larger than a substitute or competitive product. Most market potential estimates are determined using demographic data. For instance, if you are estimating the total market for a biotech product that provides blood flow to an ischemic artery immediately after cardiac arrest, you would determine the total number of heart attack victims per year, multiplied by the estimated price of your product.

Step 2: Define Your Target Market Segment

Next, you have to determine who is your best customer. This is a subset of your market potential, and represents the first group you will target, with the highest likelihood of purchasing or using your product. This is the group with the greatest need and desire for your product, and it represents the population that you must capture first to reach more of your potential market later. A Target Market is easier to reach than your entire potential market because these customers have many similar needs and behaviors. If you cannot first capture your target market segment, your product will not be successful. Define your Target Market by segmenting your Potential Market into homogeneous groups of individuals that have the greatest likelihood of purchasing or using your product. Your target market is the most homogeneous group of individuals with similar wants, needs, and desires, as it pertains to your product. Use the segmentation filters outlined earlier, such as geography, demographics, and behavior and usage, to identify your target market.

Sometimes your target market may be a niche market if it is small and specialized. In some senses, your target market can be thought of as your market niche. The truth is that you must be successful first in a niche, and seek to own that market, in order to gain acceptance from the rest of the market later. This is true of all products, including long-distance running shoes, where the phenomenally successful Nike first established a niche in the early 1970s. Since then, Nike expanded their brand to everything including the eyeglasses that I wear while writing this book. The principle of "owning" a niche, establishes your reference point, from which you later expand to the total market potential.

Step 3: Articulate Your Product's Value Proposition to Your Target Market

This is called positioning. Here you want to articulate the product's benefit to your target market. Ask yourself, "What are the product's features that infer benefit to the user?" "What are the unmet product needs of this target market?" "What are the meaningful and compelling points-of-difference between this product and the competition?" "How will the customer benefit by using this product?" For biotech products, it is useful to quantify the benefit to the user or patient, such as: a pharmacoeconomic benefit, quality-of-life benefit, survival benefit, measure of improved health, less pain, lower risk of adverse events, less complications, faster acting, etc. Start by examining competition or substitute products, and find out what value they bring to the user. Inadequate substitute product can be thought of as inferior competition, even though they are not a direct competitor since there may not be an ideal product that meets this market segment's needs. Consider the unmet needs of your customers in spite of these inadequate substitute products, and identify how your product fills these needs. Then list the solutions, features, and benefits your product provides to this target market segment.

Step 4: Select a Market Channel to Reach Your Target Market Segment

A market channel is the method you will use to gain access to your best customers, and the way you will deliver your product to them. All biotech products will have some type of market channel, and most companies will not directly sell their products to the end-user unless they are a vertically integrated pharmaceutical company.

For a simplistic example, the market channel for Rainbow® bread is the grocery stores and convenience stores. If you are Dell computer, you have more than one channel – the direct market to the end-user consumer over the internet and phone, and a market channel through retailers such as Wal-Mart and Sam's Club. If you are Ford or GM, your market channel is independent car dealers who show your vehicles to the end-user and fulfill the warranty obligations of the company.

Find ways to leverage existing market channels. Do not claim to build a new market channel as part of your market strategy, though it has been done with some biotechnology companies such as Amgen, Genentech, Genzyme, and even the newcomer Genomic Health. Ideally, you should propose to leverage existing market channels because the development of a new market channel is a new market risk. There are good market channels available to reach your target market, so pursue these rather than developing your own first. Also, look into collateral usages of existing market channels that are mutually beneficial to you and the channel partner.

Step 5: Validate and Refine Your Market Strategy

Test your market assumptions. All marketing plans are interesting theories but the only good ones are the ones that work for your product. Test your market strategy in some way to see if it is viable. The ways you can support the validation of your market strategy are many: hire a consultant, retain some physicians in this field, interview option leaders in this market, or beta test your product's receptivity. Use additional primary market research to see if it supports each of the assumptions in your market strategy. Your goal is to, in some way, validate your assumption that the target market still wants your product.

Using Biotech Product Examples

Here we take two product examples and walk through a hypothetical market strategy development process to help you see the application of these concepts and tools. Using these examples, you can draw parallels to your own product, and apply it to your own situation. The two product examples we will use are the following:

- A molecular genetic test for the assessment of breast cancer risk.
- A nasal spray biologic vaccine against influenza virus.

The market strategy you develop will be incorporated into your business plan, but nuances and modifications are refined over time. The examples here are not necessarily optimal for either of these products, but they simply serve as illustrations for application of these marketing tools.

Example: Molecular Genetic Test for the Assessment of Breast Cancer Risk

Determine Your Potential Market Size

We start by identifying the total market opportunity for this product and estimate the market size in patients and dollars. We always want to find the largest potential market for our product. For this example, the patient-customer could be both men and women because men do get breast cancer, though it is only about 1–2% of all cases. Clearly, the total potential market would double if this test were offered to *both* men and women. But to market to both men and women, you need product development milestones that include genotyping men *and* women with breast cancer in order to identify causative genes for both. However, your market research may show that today there is not an acute need for a product to assess breast cancer risk in men, so we will target only women. This reemphasizes a previous point – you do not want to have a superior technology product in search of a market need. Also, notice how market development decisions impact your product development milestones and vice versa.

We concluded that the potential market for this product is *all* women, right? Not necessarily. You need to determine how acute is the need to assess breast cancer risk in a 10-year-old or a 90-year-old female? As the need increases, the consumer/patient has a greater interest in the product. From our secondary market research, we find that in the US, annual mammography screening recommendations begins at age 40 and continues through to age 69. Therefore, it would be reasonable and logical to assume that there would be some interest in assessing breast cancer risk from women aged 40 to 69. For now, let us conclude that this is our potential market – in the US, this means approximately 65 million women, and in all developed countries it may be approximately 250 million women. If we price this test at $1,000, then the market potential for this test would be $65 billion dollars in the US and $250 billion dollars world-wide. Realize, that making a pricing decision is much more complex than presented here. Product pricing can be based on a comparison of similar technology product pricing plus a premium; it can be the amount you can reasonably charge, or it can be based on willingness-to-pay studies if it is paid by the patient. Pricing decisions will also be impacted by private insurance and government reimbursement. If you are in a country where medical care is covered by government funds, then pricing is impacted by other factors, and these criteria determine whether your product is accepted into their formulary.

Define Your Target Market Segment

We determined that our *Potential Market* world-wide is 250 million women, but this is not our *Target Market* segment. Remember the target market is the most homogeneous group that has the strongest interest in your test. You define this group by segmenting the potential market using the tools provided. Let us use demographics to further segment the Target Market. Let us say this test will be paid for by the patient until reimbursement is obtained. Therefore, a more homogeneous market segment with the desire and ability to purchase this test might be women aged 40–69, with household incomes greater than $75,000 annually. But we can still find a more interested target market segment for such a test. Secondary market research reveals that about 65–75% of all women between the ages of 40 and 69 seek mammography services annually. Women who seek mammography services and travel to these centers, want to be told they do not have breast cancer. This target market segment should be even more motivated to purchase this test than all the others. At this point we will stop segmenting because eventually you can segment too much and end up with too few customers to have a viable business. So now, we have defined our Target Market Segment for this test as those having the following demographic and behavior characteristics:

- Women
- Age 40–69
- With household incomes greater than $75,000 annually
- Women seeking mammography services annually

Segmentation by behavior and usage help us to identify and target the most homogeneous group with similar wants, needs, and desires, who would have the highest likelihood of purchasing this test. Your Target Market is the group from which you will build a market base.

Determine a Value Proposition to the Target Market

Having a genetic test to tell a woman of her risk of breast cancer is just information. But information that has no actionable plan may only provide intellectual or entertainment value. The *Value Proposition* for this test must be the ability to do something to change the outcome of your genetic predisposition. Genetic predisposition tests for Huntington's disease do not have a high value proposition because today there is little a patient can do to alter the course of that predisposition – so why would one want to know? A value proposition needs to be something so valuable to the patient, that it compels them to want to use the product. Your secondary market research reveals that if detected early, the five-year survival rate for breast cancer is 95%; whereas if detected after it has metastasized to another organ, the five-year survival rate is 21%. Secondary market research also shows that there are preventative medications available, and that women who get breast cancer, about 90%, do not have a strong family history of the disease. Therefore, one *Value Proposition* can be that genetic information coupled to action can potentially prevent breast cancer, or catch it early, where long-term survival is the greatest.

Evaluate and Select a Market Channel to Reach the Target Market

Because this product is a genetic testing service with specimen collection requirements and not a device, you will need to find an appropriate avenue to reach the patient. Since this test requires medical interpretation, you would want the market channel to involve a physician. Having a market channel encompassing all physicians would be pro-hibitively expensive to reach, and may require a pharmaceutical-type sales force. Remembering that biotechnology products have three customers, you would also segment the physician market just as you did for the patient, but the physician can also become a market channel. Your market research may show that there is movement to consolidate mammography services to comprehensive breast care centers, which have higher-end services with greater sensitivity and specificity, such as MRI and ultra-sound. Your research shows that these centers do not have a method to identify the high risk patient other than through family history, so this would be an ideal test to incorporate into their services. By targeting this market first, you have a suitable market channel and may later expand into a secondary market such as OB/GYNs and Family Practice.

Validate and Refine the Market Strategy

One way to validate the market is to do a test market of the product's receptivity. Test markets can be difficult and expensive to conduct, but they can provide good infor-mation as to patient interest and willingness-to-pay. If you have a handful of these centers with physicians interested in this test, ask them to participate with you in a limited manner. Assess the receptivity of their population to take such a test, and thus verify the physician and patient market interest. This step is necessary to be sure that all your market research and conclusions actually mirror what is going on in the real world.

Example: Nasal Spray Vaccine Against Influenza Virus

Determine Your Potential Market Size

Just as we did in the previous example, we want to find the largest potential market for this product. To do this we can consider all individuals that could be vaccinated against influenza as a starting point. Since influenza is a disease that potentially affects all individuals, the potential market is simple to estimate, which would be all individuals between certain ages. Pick a country, find the population between those ages, and multiply that number by the price of the vaccine. Let us presume that this new vaccine delivery product will have the same potential market as all other influenza vaccines. In the US, we estimate 300 million people; and in the UK, we would estimate 60 million people. If we assume a product cost of $40 per vaccination, this would be a $12 billion dollar market in the US or a $2.4 billion dollar market in the UK. Unless there are other restrictions that arise based on efficacy or safety, we would use these estimates at the potential market for a nasal spray vaccine against influenza virus.

Define Your Target Market Segment

Remember that the *potential market* is not the same as the *target market*. We want to segment the potential market population into a group we could target more specifically, that would switch from current vaccines to this product. We are seeking homogeneous groups that have similar wants and needs regarding this product. From our secondary market research, we find that the greatest risk of complications and death from influenza is in the young and the elderly. Our primary and secondary market research may also show that many of the young and elderly avoid influenza vaccination because of an aversion to getting shots. For this illustration, we will *presume* that this vaccine is equally effective for all populations as the traditional vaccine, but be aware how performance and efficacy impact your market strategy.

Now we have defined our Target Market Segment with these demographic characteristics:

- Children aged 6–12
- Older adults aged 50–80

Segmentation by behavior and usage help us identify and target the most homogeneous groups with similar wants, needs, and desires, who would have the highest likelihood of purchasing this product. Depending on your distribution channel and the value proposition to your target market, you might target the product to the entire market at the outset. Although a broad market exists for an influenza vaccine, in order to be successful you must secure strong adoption from a target market to serve as a reference for broader adoption later. A strong but targeted adoption initially, is better than any broad but weak adoption. The more strongly the target market segment or market niche wants your product, the more your product gains credibility and value, which can be effectively transferred to other market segments. Because it is usually difficult to target the largest market first, you will have to make target market choices in the beginning.

Determine Your Value Proposition to Your Target Market

Let us assume that this vaccine will be marketed at the same price point as the competitor's vaccine. A good value proposition then would be that your product provides the same effectiveness at the same cost, but with the least amount of discomfort while improving compliance. We could also price the vaccine higher than the traditional vaccines, in this case, our value proposition should just focus on the improved product effectiveness for protecting against influenza and the convenience of use. Your market research may show that resistance to influenza vaccination is influenced by pain and fear of injections. Since your product supports a painless, convenient, easy-to-administer, and effective method of vaccination, this value needs to be positioned to a target market that values these characteristics. Children and older adults, and physicians that care for these patients, are good examples of markets that could appreciate the value of your product. Also, do not forget that you have an opportunity to reach those nonusers of traditional vaccinations because now you have a different value proposition.

Evaluate and Select a Market Channel to Reach Your Target Market

Because traditional market channels exist for vaccines, you may just want to use these because they are effective and readily available. In order to preferentially carry and promote your product over traditional influenza vaccines, you may need to provide extra incentives to these distribution channels until your product gets adopted successfully. There are many ways, which we will not review, to bring attention and focus to your new product, and erode market share from other traditional vaccination products.

If we continue to follow the three customer model for biotechnology products, we also will identify and segment our physicians, and find a target market segment within this group. Because we must first focus on strong adoption by older adults and the young, we will conclude that our physician-customers are pediatricians, family practice physicians, and gerontologists.

Validate and Refine Your Market Strategy

A market can be validated by selecting a sample of pediatricians across the country, and interviewing them as to their enthusiasm about promoting such a product to their patients. You may also do this with a similar population of gerontologists. This can be done through a contract with a CRO or marketing firm. Similar market research may have been performed initially but more validation will help refine this information.

These two biotechnology product examples are used only for demonstration purposes to illustrate the application of some of these marketing tools. These market conclusions may not be ideal marketing strategies for these products, because there is so much more that is involved in developing a successful marketing strategy than what is presented in this simple exercise. However, this exercise will help you get started using these methods and tools to develop your own marketing strategy for your product.

Market Development Milestones

Market development milestones are important in demonstrating progress with your market strategy. Market development milestones serve the same purpose as your product development milestones – when reached they reduce the risk to your organization, and potentially increase its value. You will need to define market development milestones for your product just as you did for your product's development. Just recognize there is great diversity in marketing strategy, so formulate these milestones with *your* specific product in mind. Some examples of market development milestones could include the following:

- Complete a thorough review of all secondary market research to determine your target market segment needs and wants, as it relates to your product
- Complete a small primary research study on your target market segment needs
- Complete a large primary market research study using a professional organization, then refine your assumptions based on this information
- Identify your target market segment of physicians and patients or end-users for your product
- Finalize the value proposition of your product to its target market customers
- Develop a branding strategy for your product and company
- Determine product pricing
- Complete a pharmacoeconomic assessment for your product
- Determine the medical benefit in cost savings or quality-of-life improvement for your product
- Conduct primary research on willingness of private insurance to reimburse your product
- Identify suitable distribution channels for your product
- Secure private insurance reimbursement for your product
- Secure international marketing partners

Biotechnology Product Adoption Curve

Adoption of high technology products in the electronics industry provides a parallel of the challenges in biotechnology product adoption. I highly recommend the book, "Crossing the Chasm" by Geoffrey Moore, for those who are interested in learning how high-technology product adoption occurs. Although his book refers to microchip and computer industry products, the adoption curve is virtually the same for biotechnology products. The book's concept is built around the Technology Adoption Curve, which divides any target market segment into adoption stages and groups of individuals: Innovators, Early Adopters, Early Majority, Late Majority, and Laggards. Innovators are the first to use your product, and right behind them are the Early Adopters; however, the marketing chasm between them and the Early Majority is difficult to cross. Since most biotechnology products have three customers – the patient, the physician, and the payor – your marketing challenge may be threefold more difficult. Realize that even with the best market strategy, it still takes persistence, creativity, and adaptability, to penetrate a market with even the best of biotechnology products.

Summary

A marketing strategy is a carefully thought-out execution plan, based on tested assumptions that reduce the risk of marketing failure, and seeks to maximize profits. The best advice is to spend plenty of time and effort developing a solid marketing plan. Build the foundation, understand the market issues, and then seek help with refinement and evaluation. Unless the entrepreneur is an experienced marketing executive from a successful biotechnology or pharmaceutical company, he/she will need help with their marketing strategy. This is not an area of business to pinch pennies; good marketing strategy help is not cheap. There are many well-respected organizations and consulting firms that provide this type of help. Also one can call top business schools to obtain consulting help from their best marketing professors. When taking this route, be sure the professors have real life industry experience, in addition to academic credentials.

A solid marketing plan for your product will not only help raise capital but it also demonstrates that the entrepreneur understands the market issues. Venture fund managers can easily spot a shoddy marketing strategy and will walk away from a deal even if the technology is world-class. Entrepreneurs must demonstrate to investors that they have an understanding of the marketing risks, and show what they are doing to manage these risks. Be sure the plan is well thought-out and not built upon erroneous presumptions – support the plan with data and examples. The better one can affirm the marketing analogy, "will the dogs eat the dog food?" the higher the likelihood you will be successful in raising capital and marketing your product. A good marketing plan ensures that upon commercialization, the organization does not stumble in the market. The marketing strategy is an essential component of a business plan and a key factor in raising capital. Do not forget that the development of a market strategy is a parallel process that runs alongside the product development pathway discussed in Chapter 6.

Chapter 8
Financing Your Company – Part 1: Raising Money, Capital Needs, and Funding Sources

Once a company is formally established, the next step is that of raising capital – and certainly plenty of it. Raising money for a start-up company may be the single most arduous, and time-consuming activity undertaken during one's career. It may be tempting but erroneous to compare financing of a biotechnology company with those of non-life science companies such as Internet, software, or IT businesses. Financial requirements for companies in those industries are vastly different in the amount of capital required, the number of financing rounds needed, and the development time required before a product can be sold. An Internet-based company can very likely advance a product to commercialization with only a seed capital round. Whereas, for a biotechnology company, depending on the type of product, potentially three to five large funding rounds or more may be needed before a product reaches the market, and one of these funding events could be through an initial public offering (IPO), a partnership, or acquisition.

The decisions made regarding company financing impact its future growth opportunities, so it is important to understand the myriad of issues involved with financing choices. Proper financial planning is of such critical importance to the success of a biotech business that I have devoted two chapters to various topics on this subject. Even then, there is more to discuss about the funding and financing of a biotech company than can be covered in a single book. In this chapter, we discuss how much money is needed, how long it takes to find it, and where to go to seek funding. If a company carefully manages their product's development and consistently meets key development milestones in a timely manner, they will greatly improve their chances of securing the capital needed. Being properly prepared and learning some key aspects about funding will make this challenging job more rewarding. In the following section, I would like to share with you several notions that are important to understand before raising capital.

C.D. Shimasaki, *The Business of Bioscience: What Goes into Making a Biotechnology Product*, DOI 10.1007/978-1-4419-0064-7_8, © 2009 American Association of Pharmaceutical Scientists

Raising Money

Why Does It Seem Like All Our Time Is Spent Raising Money?

At some point, raising money over the course of a company's development will seem like a full-time job. Embrace this notion rather than run from it. In actuality, someone in the company will indeed be raising money continuously throughout the organization's development. Even after securing the first round of capital, the company should be preparing for follow-on financing. Raising follow-on rounds of capital is easier and occurs faster when a company continuously communicates its progress to existing and potential investors. Work on keeping the company in the public eye, and always look for opportunities to share the benefit of your product idea to potential investors. Seek opportunities for speaking at venture forums and investor meetings, irrespective of whether or not capital is immediately required. I routinely try to secure a presentation slot at many high profile investor conference meetings even when capital is not eminently needed. These types of activities can seem at the moment nonproductive, but it is building credibility and leveraging interest for later. Raising capital will be much easier during subsequent financing rounds if it is recognized that the entrepreneur's job *is* to raise money. Even after becoming successful and the company becomes profitable you will still need expansion capital to grow the business, acquire product lines or technology, make capital improvements, or acquire other companies; so do not forget that you really are, in a sense – always raising money.

Raise Money on Strength

It is easier to raise money when a company *has* money. Waiting until the company is bone dry of cash offers limited options, little negotiating power, and usually results in a desperate situation. When cash is in the bank, that is the time to begin attracting investors for future rounds. Investors do not like to invest in desperate companies; they want to invest in progressive and vibrant companies that are making exciting product development progress. Learn to raise money when money is in the bank. Remember one can always say "no" to investors, but one cannot say "no" until someone has said "yes."

Raise More Money Than the Anticipated Need

A good financing plan, is to raise more money than estimated to reach the next stage of product development. Biotech companies are guaranteed to encounter unanticipated product development delays or unanticipated technical problems. If they have raised only enough capital to reach a planned milestone, they may be forced to suspend their efforts or downsize until additional capital is found. Recovering from these types of setbacks is extremely difficult if the company does not have any cash reserves.

Another reason to raise more cash than needed is that the capital markets may not be plentiful at the time the company expects to raise money. Financing windows for biotechnology are like the windmill hole on a miniature golf course – the opportunity closes as quickly as it opens. Also, investor interest in a particular technology or market space may wane when money is required. Financing for biotechnology is also impacted by the general economy – when the economy is in a downturn, there are fewer sources of investment capital. By having more capital in the bank, a company can weather delays in financing windows such as these.

Some experts say, taking more money than a company needs presents a problem for valuation and dilution of ownership – there may be a small bit of truth in this. However, I personally have not seen this to be a problem most biotech companies face. More frequently, development stage companies are undercapitalized, left with few options, and end up running out of money at some critical juncture. In reality, company valuation problems are not based on raising more money than needed; they occur because the company consumed more money than reasonably expected to reach a key milestone, or the next stage of product development. Young biotech companies should carefully control the money spent, and be creative in economically and expeditiously achieving their product development milestones, regardless of how much money they have.

Do Not Wait Too Long to Start Raising Your Next Round of Capital

Most biotech companies operate with less than two years of cash in the bank, and more frequently these companies have less than 12 months of capital. The worst business decisions are made when anticipated funding does not arrive, and the company has little time to find alternate financing. While I was at a former start-up biotechnology company, an unfortunate situation such as this occurred. The company had successfully raised $9 million in two private placements, and another $21 million in an IPO. However, inventory was built in anticipation of sales projections for a seasonal product with a fluctuating demand that did not materialize as quickly as anticipated. Over the next three years, cash dwindled down to about $7 million, but the company had a $5 million annual burn rate. The founding CEO fired the sales and marketing staff, but shortly thereafter the CEO was fired by the venture capital (VC) investors. The Board representative for the VC investors took over as the new CEO. This individual had plenty of good business experience, but timing could not have been worse – there was a downturn in the VC market when the company needed to raise money. To compound the problem, minimal product was sold from inventory and the company was under pressure to accept a $2 million Senior Secured Convertible Debenture with a "death-spiral" feature. This meant that equity ownership was recalculated if the stock price went below certain threshold points. In the worst situation imagined, the stock price continued to nose dive – ownership continued to be recalculated – and eventually the $2 million dollar note holder became the majority shareholder of the entire organization, even though about $32 million was invested previously.

This example illustrates the point, that until a company is profitable, you must continue fundraising efforts even when it seems not necessary. Remember, external events are not under your control, and there is no certainty of finding capital when it is needed. When a company is seeking VC funding for the first time, realize that the entire process could take as long as one to two years. This includes finding a suitable partner, completing due diligence, closing a deal, and receiving funding. Allow plenty of time for fundraising, and always have back-up plans should funds not materialize by a specific date or time.

Do Not Overvalue Your Company

There is no advantage to having unusually high valuations for a company at any stage, particularly at the seed and early development stage (valuation and financing stages are discussed in more detail in Chapter 9). The higher the valuation is from reasonableness, the more difficult it will be to raise the next round of capital. Companies with unrealistic valuations, and in need of cash, usually wind up accepting a "down round." This is a situation where the previous (premoney) valuation is reduced in order to secure a new investment. In a down round, the existing shareholders lose more than just a straight proportional ownership of their equity by dilution. Learn not to be greedy or have greedy investors. When financing is done right – in the end, everyone will be rewarded. There is some truth to the saying "Pigs get fat, but hogs get slaughtered."

New Money Trumps Old Money

Each new group of investors can, by themselves, dictate the terms of the next financing round. When previous investors do not participate in the next financing round, either because they cannot or will not put up additional capital, they do not have a voice in the new investment terms. New investors are not required to maintain any arrangements old investors made with the company. This leads to another saying called the Investor's Golden Rule: He who has the gold – makes the rules.

Just Because Investors Are Interested, It Does Not Mean They Will Make an Investment

Entrepreneurs need to understand quickly that everybody is "interested" in their company and product – but that does not mean that they will put up real money to support it. There is an excitement upon finding investors who seem truly interested in the company. However, just because they are interested, this does not mean they will write a check. Sometimes investors do not have the heart to say "no," or they may have an interest in the deal, but lose interest later. Do not presume that investors are going to invest simply because they ask for additional information, or because they have a "serious interest."

Even when finding investors with real interest, and deal terms are discussed, assume that nothing is certain until a check is in the bank. I have seen some of the strangest things kill financing deals that everyone thought was certain to happen. At one development-stage company, we were raising a $5 million dollar private placement. After a lengthy period of seeking capital, one investor took a serious interest and gave their stated intent to fund the entire $5 million round – but with the condition that no other investors participate. Our proposed terms were discussed and not contested. Discussions then proceeded on how the money would be traunched in, and when the deal would close. After two to three months of further discussions, and shortly before the deal was to close, the terms suddenly changed. We received new deal terms that devaluated the company to approximately 1/8th of the current value, accompanied by extremely unfavorable terms for existing shareholders and lenders. They were unwilling to renegotiate much and we could not reach agreement – as a result we could not close the deal. Because we assumed this funding was certain, and had no back-up options, we had to scramble to get bridge capital to secure time to find other investors. This is not to advocate that entrepreneurs should be skeptical of everyone, just do not start counting money, or worse – do not start spending it before it is in the bank. A company is well-served by having more investors willing to write checks than they have need for their money. Learn to have a financing deal oversubscribed.

How Much Money Does a Company Need to Raise?

The total amount of money a company needs, depends upon the product and company business model. The amount of capital needed for a biologic or small molecule therapeutic is vastly different from that of a diagnostic, which is also different from that of a medical device. Even within therapeutics categories, costs vary; biologics can have differing costs than traditional small molecule drugs. Before raising money, determine the amount of money needed for the specific product, and estimate the amount of money needed to move through to each product development stage. When calculating the needed capital, be sure to include all expenses in addition to product development costs, such as market development and corporate development costs.

There are two pitfalls when projecting financial requirements. The first one is estimating that less money is required than really needed. This is a common problem for first-time entrepreneurs because they are not familiar with development costs, or the obstacles and types of delays that increase costs. A good way to avoid underestimation, is to base cost estimates on the amount of actual capital similar companies raised during each stage of their development, and estimate the total monies required to reach commercialization. Such information can be accessed by someone with a subscription to private equity databases such as Dow Jones VentureSource[1] or Thomson VentureXpert.[2] When comparing the amount of capital raised by other companies, be sure to include any debt capital, partnership contributions, and grant funding. To arrive at a general estimate, add the equity, debt, partnership, and grant funding and increase that by 10–20% because all situations are not alike. Confirm your total estimates by talking to other professionals involved with biotech companies to find out how much capital they raised to complete their product development stages. Knowing the total cumulative product development costs is important for estimating a hard ceiling on your ability to raise capital for products other than therapeutics and biologics. For instance, when developing a medical device, if the total development cost for the product is estimated to be $75 million dollars, and the average market valuation for acquisition of a similar medical device company is currently $150 million, the potential investment return is only 2× for the last investor round. This type of investment return would not be great enough to attract new money for that level of risk. In this situation, the company either needs to figure out how to substantially reduce development costs, or find an interested early licensing partner that would be willing to fund such product development for their own portfolio. Therapeutics typically do not have this type of development costs-valuation problem because the valuation of a successful therapeutic is so high. However, still be aware of company valuations at various stages, and make sure they are in line with convention.

The second pitfall for the biotech entrepreneur is to recognize the magnitude of the financial requirements and never begin – this is the fear of reality. Tom Perkins of Kleiner-Perkins, one of the earliest and most successful venture capitalist in biotechnology, and backer of Genentech said in a interview, "If we knew how much it would cost to fund the company, we might not even have started."[3] As it turned out Genentech required billions of dollars to reach their prominence in the biotechnology industry. The answer to "How do you raise this amount of money?" is similar to the answer to "How do you eat an elephant?" The answer is "one bite at a time." Do not be frightened or alarmed by the amount of capital ultimately needed for a start-up, because this does not need to be raised all in a single round. Just focus on raising the current round of capital needed, set reasonable development expectations, meet milestones, and communicate effectively to investors, then raising the next round of capital will be much easier.

Because multiple round of capital must be raised during product development, timing for each funding round is critical. It is easier to raise money after a key development milestone is reached because this creates an inflection point in valuation. Valuation, and the amount of money raised impacts dilution. At inception, the company has a very low valuation, and taking large amounts of capital at that stage significantly dilutes ownership. Whereas raising larger amounts of capital later at higher valuations, is not as dilutive. As a valuation strategy, make as much development

---

[1] http://www.venturesource.com
[2] http://vx.thomsonib.com/
[3] Thomas J. Perkins, "Kleiner Perkins, Venture Capital, and the Chairmanship of Genentech, 1976–1995," an oral history conducted in 2001 by Glenn E. Bugos for the Regional Oral History Office, The Bancroft Library, University of California, Berkeley, 2002.

progress as possible early with smaller rounds of capital. In Chapter 9 we review typical company funding stages and typical amounts raised.

The information below provide general ranges of capital needed for various product categories, but remember, there is variability within each of these categories; however, they provide a good first approximation of how much time and money will be needed to develop a product to reach commercialization.

Therapeutic Estimates

Drug development of therapeutics such as small molecules, biologics, or genetically-engineered drugs are by far the most costly and time-consuming category of the four – yet they provide the potential for the greatest investment return. The Pharmaceutical Manufacturers Association has estimated the costs of developing a drug from basic research to commercialization, is over $1 billion dollars and takes an average of about 12–15 years. Although these figures include the cost of failed drugs in development, any one therapeutic in development can cost in the range of a few hundred million dollars. Therapeutics and biologics, compared with the other categories discussed, have longer preclinical development and clinical testing requirements, accompanied by longer FDA review times. As an entrepreneur, you do not need to plan on raising $1 billion dollars for your start-up company because a good portion of your capital will come later through potential partnership licensing arrangements, an acquisition, or even an IPO. You must recognize that you will need potentially hundreds of millions of dollars over the life of your product development to have a chance of success in developing your therapeutic.

The biotech and pharmaceutical industry as a whole has spent much effort, thought, and strategy on reducing drug development, clinical testing, and approval times. Industry attempts at reducing drug development time have previously resulted in combinatorial chemistry platforms, high-throughput screening, and more recently in-silico models. The drug repositioning business model was developed as an attempt to reduce the time, money, and risk of new drug development by examining existing shelved drugs for new target uses based on genomic information. These approaches may shave off a few years of development time, and shorten the clinical testing time or eliminate a small amount of FDA review time, but you can still expect that it could cost hundreds of millions of dollars and take 8 to 15 years to reach commercialization.

Diagnostic Tests Cost Estimates

Diagnostic tests include, commercial test kits sold to clinical and research laboratories and physician's offices, and tests for over-the-counter use. Categories of tests include infectious diseases, metabolic tests that run on large automated platforms in hospital or clinical laboratories, and point-of-care tests run in physician's offices. The FDA categorizes these as in-vitro diagnostic (IVD) tests. The costs to develop diagnostics become higher if instrumentation and software are required to perform the test, when compared to test kits for over-the-counter or physician's office. In general, development costs for most diagnostic tests average in the range of $15–$50 million dollars. However, development costs can be as small as $7–$10 million dollars for simple follow-on tests using a previously developed testing platform. However, equipment-based testing costs can run over $50 million. For marketing in the US, keep in mind that most all diagnostic tests are required to clear the FDA before marketing, but development costs will vary based on the regulatory pathway required for clearance or approval, be it a 510(k) vs. PMA (discussed in Chapter 13). The regulatory route will affect the time to product launch, and ultimately the costs for development, clinical studies, and general overhead. For instance, a product with a 510(k) regulatory requirement will take less time, and therefore cost less than a PMA regulatory pathway because less clinical data are required. Also, the amount of information necessary for approval is less stringent, and the time to finance the company prior to revenue generation is shorter.

Medical Device Estimates

Medical devices encompass a large and diverse group of tools for medical care and treatment. Medical devices include cardiac pacemakers, catheters, defibrillation machines, X-ray machines such as mammography devices, orthopedic devices, biosensors, biomaterials, and laser surgical devices. Depending on the particular medical device and technology, development costs can range between $25 and $60 million dollars or more. Drug-eluting stents create a convergence of medical devices and therapeutics; this introduces unique challenges for regulatory review and development, and the costs for these devices are much greater than the medical device group in general. Development and clinical testing costs for new drug-eluting stents and implantable drug delivery devices may range between $50 and $120 million or more.

Clinical Laboratory Services: Genetic and Gene Testing of Molecular Markers

The discovery of new genes and expression products for predictive, prognostic, and theranostic testing is creating a genomics and genetic testing subspecialty. This category includes various genetic markers (DNA, copy number variation, epigenomic markers, SNPs, etc.) and monitors of expression levels of RNA from multiple genes. These types of tests may use complex algorithms for generating results. These tests make up the newly emerging personalized medicine arena, and provide potential for one of the most significant changes in how medicine will be practiced in the future. Regulation for genetic testing services is changing, and it appears that the FDA will add additional regulatory requirements for these tests, even though laboratory services are currently regulated under the Clinical Laboratory Improvement Amendment (CLIA). The costs to develop these services will greatly increase if FDA regulation is required in addition to the current regulation as clinical laboratory services. These types of tests may cost between $25 and $100 million dollars or more to reach commercialization. Part of the variation in development costs is attributed to the number of genes, or gene expression markers needed to generate a result. Cost variation also includes the number of individuals needed to be studied for development, clinical testing, and evaluation of performance.

How Long Does It Take to Raise Money for a Start-Up Company?

The short answer is – longer than planned. The background of the entrepreneur may often correlate with the amount of capital they can raise, and the relative speed in which they can raise it. One reason for this is that to raise such large amounts of capital, investors must have sufficient confidence that the entrepreneur is a capable businessperson, and someone who can run a successful company. If investors are not confident in the entrepreneur's skills to run and grow a successful organization, they will not invest. These are "founder effects." Figure 1 illustrates the relative time to raise multiple rounds of capital, based on the background of the entrepreneur in each category. This is a generalization and not the rule. Time is also affected by other factors such as the entrepreneur's experience, contacts in the industry, financing window, and industry subsector, so view these as central tendencies. These timeframes are depicted as relative for funding, compared with each of the other groups. Depending on the stage of the company, the actual time it may take to raise capital can be 6–12 months, to 2 years for the first institutional round. To get an estimate of how long you have to close the next round of capital, calculate the total capital in the bank, and divide this by the company's monthly burn rate to get the runway room (number of months with cash), then reduce that number by 3–6 months (your cash reserve). This is the time alloted in months to close on the next round of funding.

Academic Scientist or Physician

If an academic scientist or physician is the founder and the individual raising the capital, it will generally take them longer to attract capital than a life science industry or businessperson. Much of the longer timeframe has to do with the learning curve in attracting and raising capital. Other reasons are lack of existing financial contacts, and the time to establish business credibility. Most academic scientists do not usually start off with existing relationships in financial circles, whereas someone from a business background may already have existing venture capital contacts.

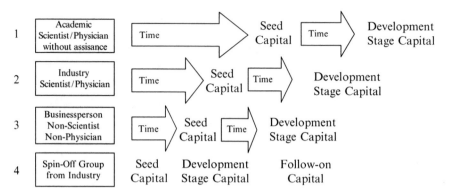

Fig. 1 Comparative time to raise capital for various entrepreneur backgrounds

Industry Scientist or Industry Physician

Another group of individuals who start biotechnology companies are industry persons with a technical background and management experience. These are individuals who are working in the biotech or pharmaceutical industry, having a Ph.D., M.D., or equivalent experience, and are in technical management positions. These individuals have the potential and can be ideally suited to lead successful biotechnology companies because they possess the technical knowledge and the business and management experience required to start and grow a biotechnology business. George Rathmann, a Ph.D., Physical Chemist and the first CEO of Amgen is such an example. Though he was not the original founder of the company, he arrived shortly afterwards, and built one of the premier biotechnology companies in the industry. Most technical entrepreneurs do not come with extensive business management experience or financing backgrounds, and this is a reason their time to raise capital is longer than a businessperson's ability to raise capital. A successful industry physician who left Genentech as VP of Clinical Affairs was Steve Sherwin, M.D., he founded Abegenix, and later became CEO of Cell Genesys.

Businessperson (Nonscientist/Nonphysician)

Many businesspeople can be successful at building a biotechnology company. The establishment of Genentech was a good example of a businessperson's ability to raise capital quickly for a start-up. Rob Swanson left his VC organization and sought out technology for a start-up, and joined forces with Herb Boyer of UCSF. The two of them created the beginning of the biotechnology industry. In this instance a businessperson joined with an academic scientist to start a biotechnology company. Each knew their strengths and weaknesses and each filled the ideal role suited to their backgrounds. Sometimes a VC group identifies a business and market opportunity based on the research of an academic scientist, and then asks them to participate in starting a biotechnology company. In this situation, the VC representative will usually be the one leading the organization. In both these examples the time to raise capital is shorter than for the Industry Scientist or Physician, and the Academic Scientist or Physician.

Splitting the technical/scientific and business/finance/market roles is common in many start-up companies, but this can create issues to be aware of, which we briefly addressed in Chapter 2. Depending on the communication and working relationship between these two individuals, this division of labor can later hinder or accelerate the growth and successful development of the organization after funding.

Spin-Off Group from Industry

Start-ups that begin by spinning off from larger companies can instantaneously raise money because of the credibility that comes with the existing staff and advanced products in development. These spin-off groups can self-form from within the larger company and voluntarily leave, or they may be motivated to leave by an impending layoff. Sometimes these spin-offs can be initiated by the parent company, such as in May 2008 when Pfizer announced the spin-out of Esperion. In this example, simultaneous with the spin-out, Esperion announced securing their first financing of $22.75 million dollars backed by four seasoned VC groups.

Major pharmaceutical companies may discontinue development of a product segment, and sometimes a group of employees will license the intellectual property and start a company. In August 2004, Baxter Healthcare Pharmaceuticals decided to discontinue the development of its respiratory therapeutics. Rather than licensing out these compounds, they spun out a biotech start-up called Aerovance with 20 employees. This start-up company began with equipment, talented employees, and intellectual property, and instantaneously raised $32 million dollars; shortly thereafter the company raised another $60 million to continue clinical studies of their mid-stage drugs. Alas, the academic scientist and physician should be so fortunate to see lightning quick capital raises as this. Unfortunately, this is not the speed for capital raises within the previous three groups.

The time horizon for raising capital can be dramatically different depending on the background of the entrepreneur. These timeframes can be months, or even years. The time to raise the first seed capital round can take the longest, followed closely by the time raising the development-stage financing, and then the first institutional round of capital. These timeframes are generalities that I have seen in practice, and are not necessarily a rule for funding; however, this will give you some idea of the patience and persistence that may be required during fundraising based upon the background of the entrepreneur.

Geographic Location Effects

Another factor that affects the time to raise capital for your biotech start-up is one you do not control – this is an address. In the US, if the address ends in San Francisco, Boston, or San Diego, a good bet is that you will have quicker success in attracting capital than your counterparts in the middle of the country, or outside a major biotech hub. Geographical location has an impact for biotech companies in all countries and not just the US. Location is not always a determinant of whether one *can* raise money, but it is certainly a factor that impacts the *time* to raise money. First some interesting generalizations:

1. In the US, the likelihood of raising significant amounts of capital is directly proportional to a company's proximity to large bodies of water. See Chapter 1 for locations of most publicly traded biotechnology companies.

2. The farther a company is inland, the longer it will take to raise institutional rounds, particularly VC.

The major biotech clusters in the US reside on the East and West Coasts near the largest sources of VC money. In Europe and the UK, there are also concentrations of biotechnology found in accepted biotech locations, and companies residing outside these locations have similar difficulties in attracting capital. When a company resides outside a biotech cluster, finding capital can be a challenge. However, this situation is progressively improving. More capital is being invested in nontraditional biotech clusters and nonhub states and municipalities are creating their own sources of funding for life sciences. In the future, it should be easier to raise capital irrespective of address, but for now, it presents an obstacle that has required some companies to relocate in order to receive a large round of capital. However, do not presume that residing in a biotech hub in any way guarantees funding.

Where to Go to Raise Capital?

There are several sources of capital for biotech companies. Each source has different interests, different limitations, and each source brings a different value to an organization. It is important to know the objectives and interests of each capital source, so you may focus your precious time on the ones that have the highest likelihood of investing. Each financing source has preferences in investing at a particular stage of development, so do not waste time by approaching them at the wrong development stage of the company. Below is a list of the most common sources of capital for biotechnology companies, accompanied by an overview of each group, and the typical stages of company development at which they invest. The various investor groups include: your own personal capital, Friends and Family, Angel Investors, Local government programs and grants, VC, Foundations, Corporate Partnerships, and Institutional Debt Financing.

Venture Deals by U.S. Region, 2001-2007

Midwest 5% | North Central 3% | Southwest 7% | Southeast 7% | Mid-Atlantic 9% | West Coast 44% | 25% Northeast

Source: Venture Capital Deal Terms Report from Dow Jones

Personal Money

Often founders put up their own money when the organization is first established, and to purchase founders shares of stock in the company. This funding gets the company formed and started. Investing one's own money demonstrates a financial commitment, and this is reassuring to the next group of investors that the founders have a vested interest in the endeavor. Having "skin in the game" speaks volumes when approaching others for financing. It says that the entrepreneur has a commitment and a strong belief in what they are doing. Entrepreneurs do not always have to put enormous amounts of their own money into a start-up company; but having some money invested early on is extremely helpful even if it is nominal. Rob Swanson and Herb Boyer each put up $500 dollars when they founded Genentech. This amount would be considered immaterial to today's investors. The question is how much cash is reasonable for a founder to have invested? Having all of your net worth invested in your idea is admirable as to a commitment, but not very smart from a financial basis. As a founder, a cash investment equal to 5% of their net assets is viewed favorably. In general, a cash investment in excess of 20% of one's net worth could be considered naïve given the risky nature of the business, and it may reflect unfavorably as to their business acumen.

If an entrepreneur does not have money to invest in their start-up organization, they should not be too concerned. It is not unusual for biotech founders not to have put up money for equity, as they will have "sweat equity" in the company. Some investments are usually tied to purchasing founders shares, which may have tax consequences if the "fair market value" of the stock was not paid. Various types of stock, and the granting and purchasing of stock options are discussed in Chapter 5.

Friends and Family

Friends, family members, and associates generally invest because they know the founders or inventors. They usually invest because they believe in them rather than because they understand the technology or market opportunity. Capital from this source usually joins or follows the entrepreneur's own money near the time of the start-up. It is a good idea to find as many other investors as possible when taking money from this group. Early stage risks are extraordinarily high, and this group usually does not have the investment experience to understand the risks, nor do they have large amounts of excess capital to lose. Though this group can be a good source of early stage seed capital, do not forget that family and friends may also be sitting across the dinner table at Thanksgiving and Christmas, so be aware of the potential for changes to these relationships.

This is a good time to talk about OPM, or "Other People's Money." You are seeking and will be spending OPM, and therefore, your responsibilities broaden to more than just the company. Because of this, it is important to have a solid plan for how much money will be needed at each stage of company development, how much money it will take to reach each value increasing milestones, and to be frugal with finite resources. Money is a precious and limited commodity to a start-up company. Money is like gas to a vehicle; you may have the most powerful engine and the most impressive make and model of automobile, but if there are no fueling stations around, you will go nowhere. Recognize that there becomes now a shared level of involvement with others, and an increased level of responsibility for OPM. Too often the founder or CEO secures funds from others, and begins to think it is their money to do with as they please. It is still OPM, and must be used and spent as such – by doing so, better decisions will be made along the way.

Peer-Reviewed Grant Funding

Many biotechnology companies fail to consider grants as a source of funding for their start-up. As previously mentioned, this is an excellent source of nondilutive funding at early development stages, and it reduces the amount of money needed to raise over the life of the organization. Most professors or scientists from an academic background are familiar with the National Institute's of Health (NIH) peer review grant system, as well as other federal agencies that have extramural grant programs. Scientists coming from industry may be more familiar with the Small Business Innovative Research (SBIR) program that awards competitive grants through 11 federal agencies. Because you are a for-profit company, your best avenue for grants would be to utilize the SBIR or the Strategic Technology Transfer Research (STTR) programs to fund some of the research and development efforts. The amount of money awarded from the SBIR and STTR can be significant in helping to build and develop the technology during its early stages. SBIR Phase I grants can provide up to

$100,000 or more for 6 months, whereas SBIR Phase II grants can provide up to $1,000,000 or more for 2–3 years. Recently, follow-on commercialization programs awarding $500,000–$3,000,000 have been introduced for successful products completing Phase I & II SBIR programs. NIH and NSF pioneered these programs that are likely to be copied by the other 11 federal agencies.

I know of an engineering technology company that has been extremely successful in writing and winning SBIR grants, such that they funded the entire company's research and development program by the sheer number of SBIR grants they won each year. Although the goal of the SBIR is to move companies from research and development to commercialization, it is possible to significantly fund R&D through grants to offset capital requirement from other sources. For those who are not proficient grant writers, you should consider the free SBIR workshops that assist in improving your grant writing skills. Alternatively, look into grant writing services that could assist in these efforts, but realize there are up-front costs and a success fee if the grant is funded. Companies that win peer-reviewed grants bring credibility and validation to their research. VCs looks favorably on companies successful in winning Phase II SBIR and STTR grants. The ability to win multiple peer-reviewed grants is an encouraging sign that the research is competitive enough for a group of peers to believe it is worth funding. There are many sources of federal grants, and each federal agency has their own program. More information can be found on the Small Business Administration website at http://www.sba.gov/SBIR/.

State and Local Financing Program Support

The bioscience industry is attractive to local municipalities because it is a high technology and clean industry that provides high paying jobs, and expands the local knowledge-based economy. As discussed in Chapter 1, there are five elements essential for building a biotech industry, and access to capital is a key. Local governments understand this and have developed programs to fund start-ups and development-stage technology companies, particularly bioscience companies, in an attempt to jump-start their local bioscience industry. Many cities and states have allocated hundreds of millions, even billions of dollars in funding initiatives for support of local biotechnology growth initiatives. Entrepreneurs should familiarize themselves with their state or local support. It is also worth contacting the local Chamber of Commerce and state Department of Commerce for more information on what programs may be available. Some of these programs provide merit-based technology loans with long-term paybacks triggered by significant funding events. Many states also have seed capital funds that coinvest with Angels in exchange for equity participation. You may be pleasantly surprised at the seed capital availabile directly from your own state. Realize that you will need as much financial help as you can get, so spend time finding out about state and local government assistance programs – it will be time well spent. Many states also have their own granting programs. If your state has a granting program, you may be eligible to make application for competitive grants, where you compete against a smaller pool of researchers. Check into your state's initiatives that support grants for diversification in new technology.

Angel Investors

An Angel Investor is a term given to high-net worth individuals who invest in early stage companies. This term is ascribed because these investors are considered "Angelic" or "heaven-sent" by coming to rescue a company at critical times, providing the life support financing needed. Most often the term Angel Investor is used synonymously with, and is interpreted to be, an "Accredited Investor." These are individuals defined by the Securities and Exchange Commission (SEC). An Accredited Investor is an individual (or individual and spouse) with a net worth of over $1 million dollars, or an individual with an annual income of $200,000 or more, or $300,000 with a spouse. They must have this income for the past two years, and a reasonable expectation of the same for the current year. An Accredited Investor is further defined by the SEC in Rule 501 of Regulation D.

Angel investors may work alone or in groups such as in Angel Networks having formal or informal communications across the country or internationally. There are some well-established sophisticated syndicate angel networks. One such example is the Tech Coast Angels that have formalized structures and funding review systems clearly outlined for potential investments. Angel investors generally invest at early stages of an organization, either during the Seed stage or Early Development (funding stages discussed in Chapter 9). Most Angel investors do not usually coinvest with later institutional investors or VC funds. A typical Angel investment can range from $25,000 to $250,000 dollars, and in rare cases an Angel investor can invest multiple millions, depending upon their interest in the technology, and their capital resources. In the US, there are over 400,000 active Angels investing in various companies. Most of the time, Angel investors can be found within one's own geographic location, and in general, they usually prefer to invest in companies within their geography. The reason for limiting an investment to their locale, is an affinity to help local companies, coupled to a lack of interest in traveling long distances to monitor an investment. However, larger, more sophisticated Angel networks will occasionally invest outside of their geography. But realistically, for a higher likelihood of success, locate Angels in your own geographic region.

Finding Angel investors can be challenging because they usually do not do this as a full-time job, and they do not advertise their activities. The best way to find Angels is by networking with people who know them, and by asking them for introductions. Check with your local university or research institution technology transfer office for names and contacts, as they may know many of the local Angels. Also check with the local Chamber of Commerce, regional economic development agency, or a technology commercialization center or equivalent – these may be good sources of Angel contacts that invest in biotechnology. Most Angels are good sources for names of other Angels. Once an Angel has invested, ask them for help and introductions to others. Having a motivated and excited Angel investor telling the company story is more effective than the entrepreneur doing it cold. There are many Angels out there – be persistent in locating them.

Angel investors are motivated to invest in companies because they can either relate to the market need, have a desire to support biotech in their community, or because they believe in the CEO and his/her ability to make it a success. At one organization I cofounded, we raised $18 million during four financing rounds from 50 Angel investors and one small local fund. Raising this amount of capital from Angels is extremely difficult because of the large numbers of investors needed and the work to manage and communicate with this number of shareholders. Also, Angels may require more time to educate since they may have limited experience and expertise in the field of biotechnology. However, there is a growing number of experienced biotechnology and high technology Angels who became successful through exits from their technology company start-ups. For us, going the Angel choice was a natural choice to get the technology developed because this company was not located within a biotech hub.

Private Placements or Private Offerings

Raising money for a biotech company is, in essence, selling underlying securities (stock or equity) in the company. As such, these activities fall under Securities and Exchange Regulations. These offering are called "Private Offerings" or "Private Placements" as opposed to "Public Offerings" such as an IPO. Specific offering documents are needed, and a good securities attorney will take care of the regulatory filings when raising capital. However, most of these activities fall under exemptions of the Securities laws. The most common exemption is called "Regulation D" and is a Safe-Harbor Exemption. Under these laws, a company may be exempt from many securities requirements, but there are still obligations and legal requirements for disclosure. Be sure your securities attorney advises the company on how to proceed and comply with any filing requirements.

When soliciting investments from individuals, they will expect to see a "Private Placement Memorandum," which is your offering document; in investor circles this is called "the book." Basically this book contains your complete business plan along with your financials, your sales projections, a potential return on investment, and a laundry list of risks associated with this endeavor. Your securities attorney will review these documents and more than likely help describe the risks, but the company will be writing the business plan. The business plan is a critical document that opens or closes the door to interests and investments in the company, so it is important to devote the time and effort to developing an excellent one (Business Plans are discussed in more detail in Chapter 9). When raising money from individual investors through a private placement, it is wise to accept money only from Accredited Investor as this assures that the investors understand the risks associated with an investment. Also, by raising money only from Accredited Investors, if the company is unsuccessful, this loss of investment does not significantly impact Accredited Investors as much as it would others. For more information about securities regulations see the SEC website.

The Venture Capital Industry

Venture Capital (VC) is composed of funds used for the purpose of investing in higher risk ventures expecting to produce higher rates of returns. In the early 1970s, a phenomenal number or computer chip and electronic industry companies were started with VC. The success and growth of these companies brought the local establishment of more VCs, which in turn fueled a myriad of companies in the Santa Clara Valley, and begat what is now known as Silicon Valley. VC continues to provide the bulk of funding for the biotechnology industry, especially during the later development stages of biotech companies where large amounts of capital are consumed.

The size of any venture fund varies greatly, from $20 million to over $5 billion dollars per fund with over 900 VC funds in existence. VC funds are managed by General Partners, and the money in these funds usually comes from individuals and institutions and even other funds. VC fund investors are called Limited Partners. VC funds are usually set up as LLCs for the protection of the Limited Partners and for certain tax treatments. There are VC funds directed to most every type of industry such as IT, Software, Life Science, Therapeutics, Diagnostics, Medical Services, and some large funds invest in all of the above. In 2006 almost $8 billion dollars was invested into biotechnology companies by VC (see Table 1).

It is tempting to ask the question, "Why all the interest in VC? Can't we continue by getting money from Angel investors?" The answer to that question is, "yes!" Some biotech companies that operate in a segment of the biotechnology industry (other than therapeutic companies) with lower development costs, may be able to build a diagnostic or clinical laboratory testing company with just Angel investors, but this is extremely difficult. Though it can be done, it is rare to find a successful biotechnology company that reaches the market without VC. Growing a biotech company without a VC requires many more individual investors, and usually these investors do not come

Table 1 Venture Capital Investment in Life Sciences by Sector for 1998, 2000, 2002, 2004, and 2006

Year	Biotechnology		Medical Devices and Equipment	
	Deals	Amount Invested	Deals	Amount Invested
1998	276	$1.5 Billion	280	$1.2 Billion
2000	348	$4.1 Billion	286	$2.4 Billion
2002	307	$3.2 Billion	233	$1.8 Billion
2004	374	$4.2 Billion	272	$1.9 Billion
2006	452	$4.6 Billion	352	$2.9 Billion

MoneyTree™ Report Q1, 2009 from Pricewaterhouse Coopers/National Venture Capital Association based on data provided by Thomson Reuters (Thomson Financial is now Thomson Reuters)

with the experience of a good VC. Before taking this alternate and less conventional funding path, first determine whether this capital source can and would be willing to sustain all of your financial needs to reach commercialization.

Venture Capital Brings More Than Just Money

A VC investment comes with much more than just cash. Good VCs come with valuable expertise, contacts, and help in guiding a business through growth, development, and market challenges. If you had a choice between $5 million dollars from a group of individual investors or $5 million dollars from a top VC firm in the industry, you should go with the VC firm even if it meant giving up more equity and Board representation. The reason for this is that the level of help, experience, and contacts that come with good VCs will improve the likelihood of success for the organization. In addition, VCs have working relationships with other VCs, and if the company runs into development or market challenges that require much larger financing rounds later, VCs can bring others to the table to help. Whereas individual investors do not typically have the capacity to sustain the level of funding required for a growing organization with expanding financial needs.

Finding the Right VC

Finding the right VC partnership is critical to the ultimate success of a biotech organization. Notice the term "partnership." Just as a company should desire to partner with the best collaborators, they must also seek partnerships with the best VCs in their industry sector. However, all VC firms are not created equal. Search for VCs working on successful deals in *your* technology or market space. Check out their track record, and research their partners and associates, backgrounds. Do they have the knowledge, expertise, and good judgment skills that would improve the decisions made by the company? Talk with VC's portfolio companies and find out if they work well with the companies currently in their fund. Do they have the patience to listen to problems, and offer sound solutions and advice? Do they have the contacts that can help grow and develop the organization? These are the types of questions to ask of VCs and their portfolio companies. When searching for a VC, start with a list of the top five to ten that would be ideally suited for the company, and then specifically target them for interest. Attracting VC interest depends on many factors such as the scientific merit of the organization, the development stage of the company, the capabilities of the company management, in addition to the number of deals the VC has already funded, and the stage of their current fund.

In general, most large VC firms do not focus on start-up companies unless they start them themselves; although this varies depending on the economy and capital markets.

Some VC funds will allocate a portion of their fund to early stage development. However, focusing all your efforts trying to attract VCs as a start-up company may be a time-consuming, low payoff, and disappointing endeavor. One reason that VCs do not typically invest in early stage start-up companies is that they have specific time horizons for their fund, and a typical fund life cycle is 10 years. It is important to ask where a particular VC fund is in their life cycle. Ideally, seek a fund that is in the first third of its life cycle because they will have the longest investment horizon. A fund in the middle of its investment cycle begins to be selective as to the stage of investment and the potential length of time to an exit. The final third of a fund's life cycle is spent predominantly on follow-up investments and liquidations in existing portfolio companies, some which are successful and some unsuccessful.

The amount of money that a VC needs to put to work is enormous. A $1 million investment in a company takes the same amount of time to follow and manage as does a $20 million dollar investment, and for a $200 million fund, the $1 million dollar investment would *not* be a wise time decision. A good use the entrepreneur's time during the company start-up stage is to inform selected VCs of your development progress, and propose to keep them apprised as you move through your product development stages. Let them know that the company is not seeking VC funding at this time, but would like to know if they would be interested in being updated on the growth and development of the organization for potential future discussions. Work to prepare future opportunities, and establish some history and credibility with these organizations.

How NOT to Find a Venture Capital Firm

Early in my entrepreneurial career while at one start-up company, I presumed that the greater the number of VCs I contacted, the better my chances of finding one that will invest. In my mind it was a numbers game. So I purchased an expensive list containing all the e-mail addresses of Life Science VC firms in the US and overseas. I then carefully crafted what I thought was a winning Executive Summary, and described the steps we were taking to turn this science into a blockbuster product. I crafted an enticing introductory e-mail, and carefully selected over 200 VC groups in the Life Science area, then confidently hit the "send" button. I presumed I would find many eager investors, so I anxiously awaited their interested responses. Out of the 200 solicitations, only 30% responded. Of those 60+ responses the vast majority of them were some version of the following examples:

- "Thank you for the information regarding your company. Unfortunately, due to capital constraints through the bank, we are pursuing very few new opportunities this year so we will not be able to pursue yours. Best of luck with your financing efforts."

- "Does not fit into present portfolio, will keep for one year and if position changes we will contact."

- "Unfortunately, we do not feel that this business would be a suitable investment for our funds and we are therefore unable to pursue this further. However, I thank you for giving us the opportunity to review this proposal and I wish you success with the fundraising."

- "After careful review we have determined that this opportunity is early for our portfolio. We typically invest in later stage technologies that have some clinical trial data. Please keep us apprised of your continued progress."

- "While we find your business proposal has merit, unfortunately it does not meet our investment criteria of the funds from which we are making new investments at this time. Should there be significant developments which you feel might warrant reexamination of your business plan at a future time, please contact us through our website or via e-mail."

- "Thanks for the information. Unfortunately, we are not interested in pursuing this opportunity because it is outside our area of geographic focus which is the mid-Atlantic region. Good luck in your fundraising efforts."

More responses continued to trickle in over the next 3–4 months, with an occasional request for clarification, or forward to a more appropriate partner in the organization. Encouragingly, I finally received seven "maybe" responses. I was excited that one of these could be my next financial partner for our new start-up company. Some correspondence continued with these seven, including an occasional phone conversation. In the end, I received one weak interest for a presentation. I flew to their office to give a presentation. I was confident that they were going to invest in our start-up company. This effort resulted in no investment. The partner I was talking with initally had only moderate interest, which was not shared by any of his partners. At the end of this 10-month process, I had no deal, and no money. After this disappointing experience, we then focused on local interest from high-net worth Angel investors and raised two rounds of capital, $1.25 million, and $2.4 million each. I share this example with you to emphasize the importance of spending time with the right VC at the appropriate time of company development, and specific to the VCs sector of interest and geographic location.

After securing a start-up and seed round, it is worthwhile to explore interest from selected VCs that like to invest in early stage development companies. It is important to have enough capital from a seed round, and a low capital consumption rate (otherwise known as "burn rate"). Make sure that the capital will at least sustain the organization for 12–18 months while exploring additional Angel and VC interests simultaneously. During this time, focus on making significant progress toward milestones to increase the likelihood that funding prospects will improve.

Additional information about VC in the US can be obtained from the National Venture Capital Association (http://www.nvca.org). VC information in the UK can be obtained from the British Venture Capital Association (http://www.bvca.co.uk); in Europe, from European Venture Capital Association (http://www.evca.com); in Canada, at Canada's Private Equity and Venture Capital Association (http://www.cvca.ca); in Germany (http://www.bvk-ev.de) and in Japan, at Japan Venture Capital Association (http://www.jvca.jp). Links to VCs in other countries can be found at http://www.nvca.org.

The Dark Side

Just as in all industries, there is a "Dark Side of the Force." Some entrepreneurs have termed these individuals as *Vulture* Capitalists, because of their aggressive opportunistic behavior, although this group comprises a very minor segment of the VC industry. A colleague shared with me an encounter with an opportunistic VC who actually believed that it was their life purpose to crush or punish the former investors as a condition of investing – just because they can – not because it was reasonable. However, this is not the mindset of the majority of venture capitalists. Just as one encounters good and bad mechanics, realtors and doctors, one will also find good and bad venture capitalists. However, do not be naive; VCs would not be in this business if they were simply purveyors of goodwill. No matter how much they like an entrepreneur, they cannot make unwise financial decisions with their institution's funds.

Seek a VC partner that plays two roles – an investor and a strategic advisor. A good VC must possess the experience and insight to add greater value to a growing business moreso than what a company could do without them. Sometimes a company does not have a choice in VC because sometimes only one avenue for financing presents itself. For this reason, it is important to start a VC search early, and target only those VC groups that fit with the organizational goals of the company. Ideally, search for a venture fund with general partners and team members that share similar goals, ideals, and values, because these individuals become partners who will be providing help, expertise, and guidance in the future direction of the company.

How Do You Approach Venture Capital?

Many VCs will say that the best way is to reach them is through an introduction from someone they know. Having an experienced biotech start-up attorney usually comes with the benefit of personal contacts to high-net worth individuals and VC firms. Your corporate attorney may be able to provide a valuable VC introduction, which is a good way to get a business plan in front of a VC firm. VCs receive numerous unsolicited business plans each year, and most get rejected. One VC partner told me that they see on average 1,000 business plans each year, then do further research on about 2 dozen. Of those 20–30 plans, they invite 10 for presentations and invest in 2–3 deals. It is easy to see that these statistics are not favorable for random deals, so it is important to have as much going for the company in order to secure VC funding. Diligently find ways to get an introduction to the VCs you are seeking, but if you still cannot find a contact, reach out to them directly anyway with key information that you believe will catch their attention.

VCs are extremely pressed for time. They usually sit on several Boards of their portfolio companies, which have needs and problems that require their help. VCs also have Limited Partners to report to and communicate the progress of their portfolio and their fund's returns. Also, VCs usually have families. With limited time and attention, be sure to cut to the chase and give VCs the most important information that is of interest to them. However, do not ignore the personal nature of a relationship, because at the end of the day, the VC partners need to like, rather than loathe working with you if you expect to have a chance of them funding a deal.

It is important to realize that VCs see many business plans containing many different technologies. Understand that anything sent to them should be considered public information or nonconfidential. For confidential information that still needs to be protected, it is best to describe briefly the information in a nonconfidential manner. Do not expect a VC to sign a confidentiality agreement to review your business plan. It is unrealistic to presume that any information sent to a VC will be held in confidence – this is just not possible to do while reviewing the massive amounts of information that they receive. Some VCs are very considerate and may actually ask permission before sending your plan or material to others. Regardless, expect that anything sent to VCs may be circulated to others from whom they may need an opinion.

Before VCs reach to an investment decision, you will be invited to meet them and give a couple of presentations to them and their partners. Face-to-face meetings are important opportunities to increase their interest in you, your technology, your market, and your company. In Chapter 4, we discussed five key elements that are critical to a successful biotechnology company. These are virtually the same elements that VCs will be looking for as they meet you and assess your opportunity. Refer to these elements and take steps to improve each one. If your business attractiveness is not great right now, a year or two later it may be a fundable VC deal. There are many books on the subject of VC funding, and it is worth the time and money to purchase and review them. Finding the right VC takes time. If you are seeking your first round of VC capital, it would not be unusual to expect a period of 1–2 years from initiation to final funding, as you search for the right VC partnership.

The Term Sheet

Successful investment interest from a VC firm will result in a "term sheet." The "term sheet" is a list of broad terms and conditions under which the VC group is willing to invest in the company. The terms outline the amount of money they will invest and under what conditions or "terms" they will invest. There are many conditions included in a term sheet, and some of the more common terms include the following:

- Type of stock: preferred, Jr. preferred, common, warrants
- Liquidation preferences
- Dividend preferences
- Redemption rights and price
- Conversion rights
- Antidilution provisions
- Rights of first refusal
- Price protection
- Voting rights
- Registration rights

The definitions of these terms are included in the Glossary. This is not an exhaustive list of terms, but it includes some of the most common ones seen in a VC term sheet. Some of these terms are negotiable and some are not. This is where your attorney will assist by helping you understand these terms and their impact on your company and future fundraising efforts. Also within the term sheet will be an implied assessment of your company valuation. This is the amount of money they estimate your company is worth, and is used to calculate the equity ownership of the new money, and amount of stock that the investor's will receive (see Chapter 9 for a discussion of Company Valuation). You can obtain an example of a standard term sheet from the National Venture Capital Association's website, in their Resources Section under "Model Legal Documents" (http://www.nvca.org).

What Is Due Diligence? How Long Does It Take?

Once a term sheet is received, the next step is for the VC group to perform due diligence on the company. Due diligence is a detailed review and thorough examination of technology issues, intellectual property, market assumptions, and corporate structure and agreements. Completing due diligence requires considerable time and effort from both the company and the interested investment group. This process can take between four and six months of work for a team and corporate counsel. Once due diligence is completed, any uncovered issues need to be remedied before closing. If successfully dealt with, a closing date will be set for funding. Closing is a joyous occasion, and the actual process is not too dissimilar to closing on a new home – just more paperwork.

The Funding Chasm: Bridging from Angel to Venture Capital

The funding stages discussed in this Chapter may appear to be presented as a steady continuum of raised capital throughout the company's development; however, this is not always the case. The two most common biotech funding groups, Angel investors and VCs, have vastly differing financing capabilities, and typically invest at different stages of a company's development. Since Angels typically invest between $25,000 and $250,000 and VCs typically invest in the range of $3–$25 million dollars or more (usually in syndicate), a potential funding gap exists for investments in the range of $1 million to as much as $5 million dollars. During the technology and market decline in 2000, these two investor groups spread further apart without much overlap, because of the huge losses sustained by VCs during the dot-com era.

The financial chasm in the funding of a life science company must be crossed to have the opportunity to reach the biotechnology Holy Grail – product commercialization. This financial chasm is the transition from Angel to institutional or VC funding. A scene from *Indiana Jones and the Last Crusade*, depicts Harrison Ford's character,

Indiana Jones passing two deadly obstacles on his way to reach the Holy Grail, only to face a third more challenging obstacle – crossing the chasm. In the movie, he closes his eyes, takes a leap of faith and a potentially life-threatening step, finding he is supported by an imperceptible surface, and safely passes the deadly chasm. This scene conveys a similar degree of potential fear and trepidation that an entrepreneur may sense, approaching this financing gap during critical stages of development. Many companies run right up to the funding edge and have no alternatives remaining. Some refer to this financing period as the "Valley of Death." It is the stage of development where a company comes closest to capital exhaustion, even though major investments have been made into the technology and product's development.

Some companies never make it past this investment chasm. Therefore, it is imperative that the company work toward doing any or all of the following prior to facing this investment chasm:

1. Find Angel Investors that have more than the typical capacity of $25,000–$250,000 to invest in your organization.

2. Attract greater numbers of Angels that have the typical investing capacity.

3. Supplement the need for Angel funding with large SBIR grant awards such as those in Phase II.

4. Keep overhead low in order to focus existing capital toward advancing the technology to reach significant funding milestones with less capital to attract VC sooner.

5. Search out VC funds that specialize in early stage development. Find those that are familiar with your technology area and have been successful in these fields in the past.

6. Plan to initiate early the VC relationship search, to jump-start the time to reach and secure investment from these groups.

7. Make use of many new transition funding opportunities available for high technology companies from state and local government.

Plan for funding gaps – especially this one. Work on as many alternatives as possible to extend your time to reach the next stage funding sources. A temporary funding gap option one may need to consider is Convertible Bridge Notes (described in Chapter 9).

Nonprofit Foundation Partnerships

A relatively new funding source for biotechnology companies has opened up from nonprofit foundations. In the past, disease foundations traditionally funded basic research through grant awards for researchers at nonprofit institutions. Recently, disease foundations have started funding drug development in biotechnology companies that are working on drugs targeting their specific disease interest. The rationale is that instead of putting all their money into basic research where the opportunity for a new drug is 12–15 years away, they can fund drug development in existing companies with drugs already in development where the potential exists to see an approved drug sooner. One example is the Cystic Fibrosis Foundation that has funded several early stage companies focused on drug development for Cystic Fibrosis. Cystic Fibrosis Foundation's Therapeutic, Inc. invested tens of millions of dollars in companies at early stages of drug development when few other investors were willing to invest. These foundations help move potential drugs one step closer to the market in exchange for a royalty on future sales. Many other foundations are following this funding model, such as the Juvenile Diabetes Research Foundation, Michael J. Fox Foundation for Parkinson's Research, Multiple Myeloma Research Foundation, Muscular Dystrophy Association and the Huntington's Disease Foundation just to name a few. If your research and development is in a disease area with a nonprofit foundation focus, it may be worthwhile to look into this type of financing assistance. More foundations are supporting biotechnology companies long-term, and this financing strategy may begin to play a more significant role of biotech drug development in the future.

Industry Corporate Partnership

Another source of significant capital comes from equity partnerships with pharmaceutical companies, or other organizations that can back a biotech company financially. These are corporate partnerships with organizations that collaterally benefit from the development of a company's product. These can be pharmaceutical companies, collateral industry benefactors, or large national laboratories. The biotech company must possess something of value to a corporate partner. It may be a product that provides increased sales of existing products, or gives them a complementary product in their existing market, or leverages their existing distribution and sales channel, or potentially something competitive to their existing products. Typically, entering into a corporate relationship with pharmaceutical companies and large industry partners occurs at later stages of product development. However, it does not hurt to solicit interest from these organizations early on, as they may have a special interest in earlier stages of development. Also they may have need for your particular product and technology to fill their product pipeline (corporate partnerships are discussed in Chapter 12).

Institutional Debt Financing

This funding avenue is one not typically available to start-up companies, but is directed toward companies having products in the market or near market introduction.

Examples of organizations that provide debt financing in the biotechnology industry include Silicon Valley Bank, Comerica, and GE Capital. These financing institutions can also structure debt for working capital backed by security on existing equipment if it is paid off and owned. These types of arrangements work at later stages of company development rather than early stages, as the risk early is very high. Occasionally other methods of debt financing are possible. Sometimes an early stage company may obtain a revolving line of credit from a local lender if they have a local high-net worth investor willing to guarantee the note. We were able to secure a $1.2 million dollar working capital loan from a local banker that was familiar with our business, and we had a high-net worth individual willing to secure the note.

Investors Expectations for Return on Investment

Because Angels typically invest in a company before VCs, they have much higher expectations on returns for their money. It is not unusual for Angels to have expectations of realizing a 30- to 40-fold return at an exit from their investment. VC investors also have very high expectations on investment returns to make their investing worthwhile. VC investments must provide the potential for at least a 10-fold return to get their attention. The reason for these high multiples in both groups is because biotech investing risk is extremely high, and many companies will not return any investment at all. To compensate for this, investors must have high expectations for returns on every investment deal they make. The magnitude of your potential investor return is something you need to be aware of as you choose your business model and estimate financial projections for your product. Another way to describe returns is using a compounded annual return rate, or internal rate of return (IRR). VC expectations are in the range of 30–40% per year during their investment in a company.

Learn to Appreciate the "Beatings"

There is a familiar story of a man who would bang his head violently against a concrete wall several times a day. When asked why he did this, his reply was "it always felt so good when I stopped." Sometimes meeting with investors and venture capitalists may feel as if you are beating your head against that same concrete wall. However, you can benefit by learning from these meetings if you are not defensive. Actively listen to their criticisms, then identify the changes that will strengthen your company and technology weaknesses. This is an important skill to learn when presenting a product opportunity – eliminate the emotion, and support your talk with facts, research and market data. VCs and investors may view a technology and future product opportunity from a different perspective than the entrepreneur;

understand their perspective because these issues will come up again in other investment meetings. Often, the problem is not that investors do not understand, but rather the entrepreneur may not have learned to appreciate other issues equally important for an investment. Learn to deal with and address these issues in future presentations, and you will have a much better opportunity for funding.

More often than not, life science entrepreneurs tend to be in love with their technology. An investor who says to the entrepreneur, "their technology and product has challenges," elicits an equivalent reaction as saying "your baby is ugly" to a new mother. Scientist entrepreneurs find their work exciting, and know the intimate details of the inner workings of the science and biology, and are anxious to "talk at" anyone who will listen. Even entrepreneurs with business backgrounds can become enamored with the science and technology because it may be what enticed them to take part in the start-up in the first place. However, be aware that this behavior produces "entrepreneurial myopia," and it becomes difficult to become objective and see the technology and product market from an unbiased view.

A great weakness of the typical biotech entrepreneur is their inability to appreciate the lack of enthusiasm from others about their technology and product. An entrepreneur's typical response to apathy is to demean their understanding, even in front of those from whom they are seeking an investment. Granted, there are individuals who just do not get it. But it is also likely that the entrepreneur is not communicating the product's value well, or possibly they do not yet have validation of a critical aspect. Rather than digging in one's heels and vigorously defending the technology application and market, acknowledge the weakness, and explain what will be done to address these areas. The entrepreneur may receive a later opportunity to share what has been done to address these concerns.

Summary

Capital for a biotechnology company comes from many sources, and one should explore all of them for potential investment in the company. Each investor group has different investment risk tolerances. Each group tends to focus on a particular stage of investing where they are familiar and comfortable with the investment risk. Seek the right ones when the appropriate stage is reached. At some point in development, VC funding will be needed. Start early and do some homework to find the ones that favor your company's technology space and business model. Be sure to target good VC partners who bring more than just money; seek a true partner who will help along the way.

Since the likelihood of raising the next round of capital is dependent upon scientific and developmental milestones made, be sure to hit most, if not all of them on time. If your organization is consistently making progress toward its development goals, and these milestones are significant, it will be easier to continually attract the needed capital for the company at all stages. Once the company has been successful in raising capital, be extremely efficient in managing capital irrespective of how much is raised. Efficient usage of capital is one hallmark for excellent returns in biotechnology companies, and this correlates with better performance in this industry. Learn to efficiently use capital as if it were sparse, keeping your sights set on your product, market and business development milestones. Ensure that each step taken in your product development path improves, and does not inhibit, the ability to raise a subsequent round of capital. Be sure that development milestones are timed to be completed before the next round of funding is needed.

A sad but true fact is that there is just not enough capital in the market to go around to all the good ideas for biotechnology companies. This means that many good ideas will not get funded for various reasons. Securing biotech funding requires persistence, and the ability to learn from each investor meeting to improve the chances of funding the company at subsequent junctures. The greatest idea imagined, the most effective drug ever conceived, or the greatest life-saving medical device dreamed of, is of no consequence if one cannot finance its development to commercialization. Someone once said "a vision without execution is a hallucination." To have a vision without funds to execute it is an exercise in frustration and futility. Persistence, flexibility, creativity, and working with exceptional people are key ingredients to successful fundraising.

Chapter 9
Financing the Company – Part 2: Funding Stages, Valuation, and Funding Tools

In the previous chapter, we discussed the consequences of overvaluing a company at the time of funding. However, company value can decrease from a previous round, even though the company was not overvalued at that time. This situation can occur when a company consumes massive amounts of money without reaching any new value-enhancing milestones. A biotech company's value is closely tied to its product development progress; therefore, it is important to understand investor expectations at different financing stages. In this chapter, we review the typical funding stages for a biotech company, discuss valuations and how they are calculated, and review typical exit strategies for a company. Practical guidelines are also presented for writing a business plan, and tips are provided on making effective presentations to potential investors.

Financing Stages for a Life Science Company

Capital for a company is raised in distinct stages that correlate with a product's development and commercialization status. Although financing stage terminology can vary, these are some of the most common usage of terms. Some companies go through each of these funding stages, and others move quickly into later stages of financing. It is worthwhile for the entrepreneur to be familiar with each stage of financing.

Start-Up or Pre-Seed Capital

Start-up Capital may also be called Formation Capital or Pre-Seed Capital. This is usually the smallest amount of capital raised, and typically used to establish and form the corporation or business entity. This can be as little as $1,000 or as much as $250,000. Money at this stage usually comes from the entrepreneurs, friends and family, and occasionally, loans secured against personal assets of the entrepreneurs. Start-up capital usually does not allow the company to accomplish much more than

C.D. Shimasaki, *The Business of Bioscience: What Goes into Making a Biotechnology Product*, DOI 10.1007/978-1-4419-0064-7_9,
© 2009 American Association of Pharmaceutical Scientists

form the organization, get some business cards, help create the business plan, and establish an entity to accept the intellectual property or technology license. If the company is fortunate to have state and local funding opportunities, they may find suitable programs that provide start-up financial assistance as discussed in the previous chapter. If a significant amount of money is raised at this stage, then there is no distinction between Start-up Capital and the subsequent Seed Capital stage.

Seed Capital

Seed Capital may be called Proof-of-Concept Capital, and is the first significant round of capital raised. This money allows the company to move the business forward, initiate or advance product development, expand the market research, hire some consultants, and in some cases, hire the first employee. This round can range from $100,000, to over a million dollars. Money raised during this stage can come from the entrepreneurs, friends and family, and Angel Investors and local funding programs supporting life science development. In addition, capital may come from research grants. Money raised during this stage is understood to be very high risk.

Early Stage Capital: Series A/B Preferred Rounds

This is a significant funding round for the organization. Many times this money comes from Angels, although it could be the first institutional investment from VCs that have a focus on early stage investing. It is not unusual for these rounds to range from a million dollars, up to tens of millions of dollars depending upon whether the product is a therapeutic or a diagnostic or medical device. Money invested at this stage comes with certain preferences that are above all other investments made in the company. Shares issued to early stage capital investors are called Preferred shares, which bear certain rights discussed in Chapter 8 under "The Term Sheet". Early Stage Capital rounds are considered to be Series A Preferred and Series B Preferred Rounds. Financing rounds from here forward are labeled alphabetically.

Mid-Stage or Development-Stage Capital: Series C/D Preferred Rounds

These are follow-on rounds that usually involve some or all of the investors in the previous preferred rounds, plus new investors. Many more VCs invest in Mid Stage Development companies than in Early Stage Development companies. The number of rounds and size of these rounds vary, and the C, D designation is only an example.

If the company's product has progressed far enough in development, sometimes a Series B Preferred Round can be considered a "Mid Stage Round." This may occur if the product is either a molecular test, diagnostic, or simple medical device, where the capital needs for product development are relatively small, and the time for development is relatively short. These rounds typically are in the tens of millions of dollars.

Later-Stage and Expansion Capital: Series E/F Preferred Rounds

Therapeutics and biologics require more rounds of capital and larger investments during each round when compared with diagnostics, medical devices, and molecular tests. More VC money becomes available as companies moves into later stage development. For therapeutics and biologics, at this stage the company is usually in human clinical trials, such as phase II or even phase III, and potentially for more than one indication-for-use. Some very complex genetic expression tests and combination medical devices may also require extensive human clinical testing and require later stage capital to reach commercialization. Preferred Series designations can continue alphabetically such as G, H, etc.

Mezzanine Capital

This is usually the last round of capital needed before an IPO or an exit for the investors. Venture capital funds are plentiful at this stage when product development risks have been greatly reduced and larger amounts of capital can be put to work. Investors at this stage enjoy a shorter time from investment to exit than for those who invested at early or development stages. Consequently, the returns on investments are lower at this stage than those for those investing at earlier stages of company development. For this reason, VC funds may balance their portfolio with some early, later, and mezzanine capital investments.

Initial Public Offering or Acquisition

For biotechnology companies, particularly drug development and biologics companies, the requirement for capital is so high that an IPO really becomes a later stage financing event rather than simply an exit for investors, although it does accomplish both (Exit strategy is discussed later in this chapter). Given the tremendous amount of capital required for drug development – up to $1 billion dollars, and the length of time to reach the market – up to 15 years, it is not feasible for any group of investors to fund the company from start-up to commercialization without a pubic financing event or significant partnership or acquisition. Large amounts of money can be raised in an IPO with the availability of follow-on financing in the public market. This is what makes an IPO an attractive later stage funding event for the company and its venture capital investors.

Funding Gaps: Convertible Notes or Convertible Bridge Loans

Occasionally, there are gaps between funding events. A start-up or development stage company can be very close to running out of money, but still be making significant product development progress, and generating interest from investors. One financing instrument used to close this funding gap is a Convertible Bridge Note or Bridge Loan. These sometimes fill the funding chasm between Angel funding and Venture Capital described in the previous chapter.

Convertible bridge notes are loans that bear a risk-appropriate interest rate and a right to convert the principle and interest into the next financing round, at the same terms the next investors determine. Usually, convertible notes also provide an extra incentive to the investor in the form of stock warrants. If warrants are provided, they are based on a percentage of the amount of the note, which can range from an additional 20–40% coverage depending on what it takes to get investor's interest. The convertible note avoids having to make a valuation determination for the company, and provides its investors with the same benefits that the larger investors will command later when they invest. A company may have interested investors, but when these investors cannot put up the amount of capital needed to really make significant progress, convertible notes provide a way to support the company, and buy time to bring in larger investments. Convertible notes are not an ideal funding mechanism and should be avoided if at all possible. A company usually considers convertible notes because they do not have much time before they run out of money, and may not yet have commitments from institutional investors who can provide significant funding to make the advancement required.

What Is the Typical Sequence of Funding Events?

There is no magic formula or universal funding route for all life science companies to follow. However, one route may look something similar to the following:

1. **Start-up Capital** may be provided by Founders, Friends and Family.

2. **Seed Capital** may be provided by Founders, Friends and Family, and Local Government entities, and Grants directed toward funding the early R&D.

3. **Series A Preferred Capital** may be provided by Angel Investors and/or Local Government Entities, Disease Foundations and possibly early stage Venture Capital and Grants directed toward funding the early R&D.

4. **Series B Preferred Capital** may be provided by Angel Investors, Disease Foundations, Syndicated Venture Capital, and possibly Corporate Partnerships.

5. **Series C Capital and Later Stage Capital** may be provided by Syndicated Venture Capital and Corporate Partnerships.

6. **Mezzanine Capital** may be provided by Venture Capital and Large Financing Institutions or Investment Bankers.

7. **Acquisition or Strategic Partnership** with a large corporation brings capital from the larger partner or acquiring company. Alternatively, an *Initial Public Offering (IPO)* brings capital that comes from the public and institutional investors purchasing shares in the open public market.

It is highly improbable for a biotechnology company focused on therapeutic development, to go from friends and family funding, to angel investors, to an IPO or acquisition without the financial help of a venture capital firm. However, it could be possible for a diagnostic company to go from funding by friends and family, to angels, to an IPO, corporate partnership or acquisition, but this depends upon the financial strength of the angels, and the amount of capital needed to get the product to the market.

How Much Money is Raised at Each Funding Stage?

The average amount of capital raised during each funding stage can vary greatly within each sector of the biotechnology industry. The amount of money that is raised during any stage is influenced by:

1. **The type of product.** Whether the product is a therapeutic, biologic, vaccine, medical diagnostic, medical device, point-of-care test, or clinical laboratory test.

2. **The financial market's interest in a disease segment or technology area at the time of raising capital.** There are phases of investor interest for disease segments. In the past, antisense DNA was very popular as were antisepsis therapeutics, each being funded relatively quickly. Now however, it is hard to find much interest or high valuations for these types of companies. More recently, genomics, proteomics, and RNAi companies have generated high interest and have raised great amounts of money.

3. **The strength of the IPO and acquisition market for biotechnology companies at the time of raising capital.** When market valuations as a whole increase or decline, it impacts the amount of money that can be raised for any development stage company. During 1999 and 2000, valuations for all biotechnology companies significantly increased because the IPO market was strong. Exits were available for investors providing excellent returns, and many companies raised significant amounts of capital. Later, the market downturn reduced valuations of early stage companies because the exits were not returning the profits investors expected, and funding rounds became much smaller and more difficult to secure.

Table 1 Biotech funding stages, estimated valuation ranges and amounts raised

Stages	Product characterization	Valuation ranges	Amounts raised	Funding sources
Start-up (pre-seed)	Concept	$1–$3 MM	$1,000–$25,000	**Entrepreneur/friends and family/ supported by: SBIR/STTR/local grants/loans**
Seed	Proof-of-concept	$2–$5 MM	$0.10–$1 MM	Entrepreneur/friends and family/Angels/ some VC/supported by: SBIR/STTR/local grants/institutions
Early and Development stage	Development	$3–$25 MM or more	$1–$20 MM multiple rounds of same	Angels/VCs/private equity/institutions/ supported by: SBIR/ STTR/local grants/ institutions
Later and Expansion stage	Development/ clinical testing	$10–$100 MM or more	$5–25 MM multiple rounds of same	VCs/private equity Institutions supported by: SBIR/STTR/local grants/institutions
Mezzanine	Market launch	$25–$100 MM or more	$10–$50 MM	VCs/investment banks/ private equity/ institutions

The amount of capital raised at any stage varies greatly and depends on the biotech segment, product type, stage of product development, and company business model. Large variations also exist in the amount of capital raised per round. To get a more accurate idea of the range of capital, look into the amount of capital secured at each round for companies comparable to yours. General funding ranges at various funding stages are outlined in Table 1 (Companies developing therapeutics and biologics may be at the higher end or above these ranges, whereas, diagnostic and medical device companies may be toward the lower end of these ranges).

How Is Company Value Determined?

Financial value is ascribed to a company prior to and immediately after funding events. This is referred to as the company valuation. A broad range of valuations exist for companies in different segments of the biotechnology industry. For instance, valuation at a Seed Round for a diagnostic company could be $2 million dollars, whereas valuation for a therapeutic company at a Seed Round could be $10 million, and in another therapeutic segment it may be $5 million.

Valuation can be – but should not be – one of the most difficult and contentious areas for a biotech entrepreneur. Company valuation is important to all parties because it impacts the amount of equity ownership given up to new investors, which in turn determines the change in ownership for the current investors and founders.

Since no group of investors can own 135% of a company, each time new money is raised, ownership percentage must be reallocated.

Valuation plays a central role in how the ownership pie is sliced – the greater the valuation, the smaller the portion given to new investors. Two terms to know are, "premoney" and "postmoney" valuation. Premoney valuation is the value ascribed to the company *before* a new investment. Postmoney valuation is the premoney valuation *plus* the value of the new investment. For instance, if a start-up company has a $2 million dollar valuation, and they raise $3 million dollars, the postmoney valuation is $5 million.

Postmoney Valuation
$2 Million + $3 Million = $5 Million Dollars

The amount of equity sold to new investors is based on the company's current valuation. If a company is raising an Early Stage Series B Preferred round and their premoney valuation is $15 million dollar, and they raise $8 million dollars, the amount of equity given to the new investors is calculated as follows:

Percentage of Company Sold to New Investors
at $15 Million Premoney
$8 Million/($8 Million + $15 Million) = 35%

Therefore, the new investors will receive 35%, and the previous investors will hold 65% of the company instead of 100%, but at a higher valuation. Existing investors are diluted when new money is raised. However, once significant product development progress is made, it can provide an inflection point in the company valuation. Using the same example, if the company valuation increased to $30 million (vs. $15 million) then less of the company is sold to the new investors for the same $8 million dollars.

Percentage of Company Sold to New Investors
at $30 Million Premoney
$8 Million/($8 Million + $30 Million) = 21%

Dilution is minimized as valuation is increased. Instead of giving new investors 35% ownership, the company now gives up 21%, simply by making value enhancing progress between funding rounds. This is why it is important to carefully plan and meet product development milestones. A good financing strategy is to raise less money at earlier stages, provided the company can advance the product far enough and increase the valuation significantly by the time larger amounts of capital are needed. However, do not attempt product development on a shoe-string budget, and do not forget that a company may not be able to raise capital when it needs the money. Balance the company's ability to make significant product development progress, with the timing and frequency expectations of each financing round.

Before we discuss valuation methods, it is worth reiterating something discussed at the beginning of Chapter 8: *do not* overvalue a company at any stage of development. If the first statement is not clear – *do not* overvalue the company! There is a temptation to significantly increase the company's valuation after each round of capital raised. This is quite reasonable *if* the company is making indisputable progress *and* the valuation is comparable to valuations of similar staged companies in the industry. However, if a company's valuation becomes significantly higher than comparable valuations at its stage of development, there are two risks:

1. A high valuation turns away potential investors including Venture Capital. They may see the deal as too rich, or they may not want to go through the trouble to restructure a deal.

2. The company has to take a "down round," with a lowered valuation. When this happens the existing investors get crushed and nobody is happy.

Artificially high valuation problems usually occur more often in early stage and development stage companies before VC investors join. It is foolish to maximize the short-term value of the company only to penalize founders and shareholders later. Remember all this is "paper money" until there is a liquidity event where shareholders can exit or sell their shares.

Valuation Methods

Numerous valuation methods exist to estimate the value of well-established corporations. These formulas are readily available and relatively easy to calculate. The trouble with these methods when applied to a development stage biotechnology company is, that these methods use financial metrics such as earnings, revenue, sales, EBITDA, or earnings per share. A development stage biotechnology company can be said to be "nonprofit" – but not by design. An alternate valuation method could be used such as Discounted Cash Flow, which takes projected annual sales and calculates the net-present-value of the future cash flows, discounted with a risk-appropriate cost-of-capital. The details of this method are not relevant for discussion here, except to understand that estimating sales of an undeveloped biotechnology product may be like throwing darts on a wall – everyone hits a different number. Few investors favor any of these methods for determining valuations of development stage biotechnology companies.

Development stage biotechnology company valuations are really driven by VC-determined values because they are the largest financers of later stage biotechnology companies. VC-determined valuations are driven by the return they can get for these companies upon exit such as an IPO or acquisition. Venture capital requires a certain return-on-investment to maintain a successful portfolio. Most VCs require the opportunity for a 10× return on their investment in 4–5 years, so the lower the exit values, the lower the valuations they ascribe to development stage companies. If exit valuations for a biotech segment, say medical device companies, are currently fetching $200 million, then the highest valuation a VC could possibly accept for a later-stage medical device company would be approximately $20 million.

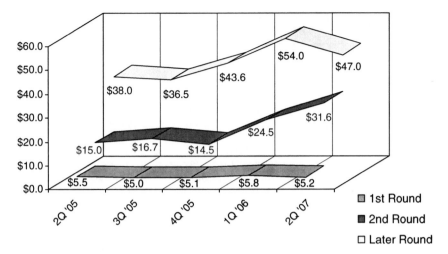

Fig. 1 Median company valuations by round and over time
Source: Venture Capital Deal Terms Report from Dow Jones

As valuations and exit prices change, so do valuations for development stage biotechnology companies. In practice, VCs are aware of conventional valuation ranges for each biotech segment and their corresponding development stage. VCs tend to invest and stay within those ranges.

Another way to arrive at a valuation for a development stage biotech company is through the use of comparables to similar companies within the same segment and at the same development stage. For an idea of how company valuations at some stages can vary over time, the above graph represents the median premoney valuation by round class from 2Q '05 to 2Q '06 for US venture-backed companies (Fig. 1).

How Much Will the Founders Own?

Unlike a family-owned business where a founder can expect to retain majority ownership of the organization, building a biotechnology company is a group project. Therefore, the founder needs to realize that by the time the company is ready for an exit, they will own a minor portion of the organization. This ownership can range from 5% to 15% depending on the amount of capital raised to reach an exit. Do not be too discouraged – 5% of hundreds of millions is much better than 100% of nothing. Table 2 shows typical founder's ownership during the early and later stages of fundraising.

It is not unusual for a founder to end up with between 5 and 15% ownership of the organization at the end, but this also depends on which segment of the industry they occupy. This does not mean a founder needs to readily give up equity in the company. The strategy is to get as much accomplished with as little early money as possible, then reach significant milestones to increase valuation, which will reduce the equity given up by all stakeholders in subsequent rounds.

Table 2 Percentages of Founders' Ownership After Round

	1st Round	2nd Round	3rd and Later
Mean	41%	24%	15%
Median	40%	20%	10%
Minimum	0%	0%	0%
Maximum	77%	84%	70%
25th Percentile	30%	12%	5%
75th Percentile	52%	36%	18%
Total Respondents	80	52	55

Source: 2009 Venture Capital Deal Terms Report from Dow Jones.
Figures rounded. Based on financings closed from July 2007 through
June 2008

Table 3 Percentages of Company Sold On a Fully Diluted Basis

	1st Round	2nd Round	3rd and Later
Mean	43%	29%	26%
Median	40%	31%	25%
Minimum	14%	1%	2%
Maximum	100%	60%	75%
25th Percentile	30%	20%	15%
75th Percentile	51%	37%	33%
Total Respondents	80	51	56

Source: 2009 Venture Capital Deal Terms Report from Dow Jones.
Figures rounded. Based on financings closed from July 2007 through
June 2008

How Much of the Company Is Given Up in Each Round?

Table 3 shows the mean, median, minimum, and maximum percentages of the company sold on a fully-diluted basis in each round (fully-diluted basis means all stock outstanding plus any options and warrants issued or granted as if they all were converted into shares). Based upon this information, most companies are giving up approximately 1/4 to 1/3 of the company each time they raise new money. Giving up 1/3 of the company three times does not mean that there is nothing left after three rounds. Remember, ownership gets recalculated each time for everyone, so investors in a previous round also get diluted if they do not participate in the next round.

Communicating to Your Shareholders

Once a company has shareholders, they need to keep them informed. Communications with shareholders on a regular basis is vital for them to stay knowledgeable about company progress and the issues facing the organization. This is important for a number of reasons:

- Existing shareholders can be the best source of locating new investors for additional rounds of financing if they are happy with the progress of the company.

- Existing shareholders can consume lots of time if they must be informed individually, and it is much easier to proactively share information with all of them simultaneously.

- Poor communications with shareholders create problems when adverse events occur. Consequently, any corrective action a company may want to take that requires shareholder approval may be difficult to obtain.

- Well-informed shareholders more readily support the company and their decisions. The frequency of communication with shareholders may be variable depending upon the type of shareholders and the stage of company development. Incorporated companies usually have articles of incorporation outlining the frequency and timing of shareholder meetings. However, in the beginning, the Board of Directors may constitute all, or the majority of the company shareholders.

I have seen start-up companies with less than five employees hold annual shareholder meetings for an entire day discussing plans, progress, and their financial condition. They open this meeting to selected guests, prospective future shareholders, and friends. At one development-stage company when all of our investors were angel investors, we sent quarterly or semiannual shareholder update letters along with newspaper articles, press releases, and CDs containing any television news coverage to keep the shareholders informed of company progress. The forum and frequency of shareholder updates and communications should be based on the number of shareholders, their involvement in your organization and the amount of money invested (the larger the amount, the more involvement and communication that is usually required). Whomever the investors, it is always good to maintain proactive communications with all of them. The company may need investor support in the future, and they will be more willing to help if they are kept well-informed.

The Value of Public Relations in Raising Money

It is amazing how effective, properly placed, and strategically timed Public Relations (PR) is in helping to raise capital. Public Relations encompass many activities and it is an effective means to highlight a company and create awareness. Good PR accomplishes this through press releases, media attention, and other types of public activities. PR is an important tool to use heavily during the development stage of a company.

Think about it – if someone sees a company and its founder on the evening news, or reads about them in the local newspaper, there is a sense of credibility and validation that accompanies media attention. Let us say a founder is presenting to a group of angel investors in a city, and someone says, "Hey, I think I saw you and your company on the news last week, what you are doing sound exciting!" Do you think PR may have made a slight positive impact on the ability to raise money from these investors?

How about another situation when presenting to a similar group of investors, and someone asks "How long have you been in this city, I don't believe I have ever heard of you or your company before." Is it possible that these two investor meetings could have potentially different outcomes? The truth is, we are all influenced by many things when making decisions, especially investment decisions. An investment decision is affected by the investor's familiarity with the technology, the market need, and the individuals leading the company. But all other things being equal, having your company in the local evening news, and written up in the local newspaper, makes it easier to raise money. We all know that the news media is not a surrogate for technical and clinical validation of anything, but it projects a sense of success and brings a measure of celebrity status. However, do not take this notion too far, because being in the business section or on the front page can be famous or infamous – both get attention.

Now that we have established there is value in getting into the news media, how does one go about doing this? First, it is important to have public relations help. Whether a company does this themselves or hires someone on a consulting or contract basis, they need to have a plan. Start by creating a press release strategy timed upon reaching significant milestones for the company, and their product's development. Then select a press release template that can be used to add content when sending to a city's major newspaper. Most local papers are looking for interesting stories about local technology companies and their future products. Examples and templates of press releases can be found on all publicly traded, and many privately held biotechnology company's websites.

Early stage biotech companies benefit from good public attention, but do *not* send out press releases for insignificant events. For example, do not send a press release to a newspaper announcing that the company launched a new website. By sending press releases with boring or insignificant information, newspaper reporters learn to ignore the company and their announcements. Some examples of subjects for interesting press releases include, the establishment and seed funding of your company, formation of product development partnerships, prototype testing, and raising of capital. These subjects could also be made boring if they are not positioned well for the readership. Remind the public of the company mission and that the company seeks to provide products or services to better the human condition, or improve the health and well being of individuals. Be sure that the company purpose is loud and clear in press releases. By providing press releases that appeal to a newspaper's readership, reporters will look forward to receiving additional information in the future. It is also good to get to know the local business reporters. Most local television stations take queues from the newspaper and follow-up with news segments. Good newspaper coverage will lead to local TV news coverage, which in turn may lead to a news syndicated station picking up the story, or even national coverage. If a company is unsuccessful in getting its press releases printed in the city's major newspaper, try a smaller local paper first. Many times they are looking for human interest stories, and may pick up this information more readily.

For help, find a local PR firm that would be willing to work with the company during its early development and one that believes in what the company is doing. Search for one that will pay attention to the company at their particular stage. Depending on the amount of work provided by the PR firm, a company may want to grant a small portion of stock options for their services as additional motivation for the long-term. At one start-up company, we were very fortunate to work with a local PR group run by an individual with a team that shares similar values, and they had experience in taking stories from the local level all the way to national news coverage such as CNN, the Today Show, and other national spotlight venues. From the formative stage they worked with us, including during difficult financial times, providing us services even when we were unable pay for them at the time of service. They came with strategy experience at the national and local level, and wisely employed former news anchors who knew the requirements for a good news story, and were able to get us into the evening news at strategic times of our development. This aspect of the business was critical in our fundraising efforts, and more so later during our sales and marketing efforts. I like to view PR in the context of the philosophical question "If a tree fell in the woods and no one was around to hear, did it really make a sound?" A corollary to that is "If you are doing something of medical importance but there is no news, it is difficult to raise money because no one has heard."

Exit Planning: Sooner not Later

An exit strategy is the method and plan for how investors will exit and be rewarded for their investment in the company. It is essential to have a good idea about an exit strategy before raising money and writing a company business plan. This may seem a bit premature – but to investors it is not. Although entrepreneurs may plan on being in their business for the long haul, investors are usually not in it for the same reasons. Yes, investors may believe in the plans to improve the health and well-being of individuals. But they are investors, and expect one day to receive a fair return that is much better than putting their money in a certificate of deposit or mutual fund. Investors need to know the company's exit strategy, and the type of return possible prior to investing. Angel investors may, or may not be as familiar with the types of exits for the biotech sector, so this information is especially important to provide to them.

Investors invest for returns. They may be familiar with the odds and risk of an investment, but if the returns are not attractive enough – they will not invest. This is also true of Venture Capital funds. However, VC professionals will know the potential returns and the potential exits better than the entrepreneur. This is why VCs have unwritten investing rules, such as how much money can go into any one deal, or what is a maximum valuation for a company in order to achieve the returns they are seeking.

What Are the Exits?

Companies in other industries can plan on growing to profitability *and* providing their shareholder's with dividends – this is not a realistic business strategy you can expect from a start-up biotech company today. The exit options in the biotech industry are:

1. Public Offering
2. Merger or Acquisition

The attractiveness for a biotech IPO depends on many things, such as the general market conditions for IPOs, investor interest in the high technology sector, the success of other biotechnology companies in a similar sector, and the overall economy. Be aware that an IPO for a company will also bring onerous financial compliance and additional governance requirements. Valuation risks to the a company ratchet up as a company becomes public and subject to the same scrutiny that highly profitable Fortune 500 companies face daily. Additionally, a public company will also have to deal with the fluctuation of public interest and stock valuation pressure.

Although an IPO is glamorous, IPO's can sometimes be difficult to reach depending on the favorability of this financial window. There are also costs of going public, it is estimated that about 15–30% of gross proceeds are consumed when a company goes public.[1] In addition, there are added expenses of compliance with Sarbanes-Oxley, estimated to cost between $1million and $2.5 million dollars. When the IPO market is good, it is possible for biotech companies to get their investors the 10× return typically sought by venture capital investors. However, when the IPO market is soft, investors may only see returns in the 2–3× range. There are additional ways to become publicly traded, such as reverse mergers with another publicly traded entity or shell. A reverse merger is less expensive than an IPO, and is relatively easy for a company to accomplish once they find a suitable shell company in which to reverse merge. This transaction then provides liquidity for shareholders as their shares would be exchanged for the publicly traded shares of the joined company. Some investors believe that mergers and acquisitions provide a better avenue for an exit and better valuations than an IPO, but again, it depends on the market interest in your company and its products and the strength of the IPO financing window. Spend time researching the M&A and IPO markets for comparable companies. Get to know the valuations received for these companies. Learn to calculate a potential return for investors by knowing what comparable company values are at each type of exit.

The Tools

The Business Plan

It is essential to put together a strong business plan to have any hope of raising the capital needed to develop a product. It may be possible during the start-up stage to get a company going with a shoddy business plan; however, when working to raise millions of dollars, a business plan will either open, or close doors. If the company is just being established, start working on the business plan now. If a business plan is already written, reexamine it to see if all the essential elements are covered. Do not send out incomplete business plans because it communicates naiveté about an entrepreneur's understanding of what is important for biotech business development.

---

[1] Pricewaterhouse Coopers, "Growing Your Business: Is Going Public Worth it?", March/April 2005.

A business plan should be updated regularly. If significant changes have occurred in a company's scientific approach, or modifications have been made to a marketing strategy, or changes have been made in a business model, update the business plan as frequently as warranted. Regularly, does not mean every two years. Do not let a potential investor read an outdated business plan and then ask questions that have no relevance to your business now. Always date each revision of a business plan.

The entire business plan should not exceed 30 pages in length. A business plan must contain a table of contents and an Executive Summary. The Executive Summary integrates all the elements of the business plan, ideally no more than three pages but it can be as many as five. The Executive Summary must provide enough information to stand alone and entice investors, such that they will want to read the business plan. It succinctly summarizes everything for the reader but they know they can go to the business plan for more detail if desired. Frequently, investors and VCs will only ask for an Executive Summary, and depending on their interest, they may or may not request anything else. The Executive Summary should be written last, and it may be the hardest to compose because it needs to be concise yet complete – the hardest part is knowing what to include and what to exclude.

A good business plan will contain the following information described below. The order presented here builds upon a storyline that makes sense to read; however, the information may be presented in any order. Many of the following sections can be combined, and the length of each is variable, but each section should be addressed.

1. Cover Page

This may seem obvious but the cover page of a business plan should contain all the necessary contact information to reach the entrepreneur, and should also show the plan revision date. Some companies number their business plans, and for private placement purposes, they would keep a record of who received them – consult your corporate attorney for issues with tracking them.

2. Company Focus, Mission, or Purpose

This portion tells what the company does. It provides a short overview and summary of the technology or product and the value it has to the world. Summarize the company and its focus in a way that does not use trite phrases or buzz words that attempt to hype the business. Overused words like, "cutting-edge," "world's first," and "one-of-a-kind" do not help in impressing sophisticated investors.

3. Company Background

This section does not have to be lengthy. Expound on what the company does, its location, history, and any important milestones achieved by the company. This may be combined with the company focus, and presented as an overview and background.

4. Market Opportunity

This section is critically important because without first establishing there is an unmet market need for the product, there will be no investment interest. It does not matter how great the technology if the reader is not convinced there is a need for the product. This is done by describing the problem and the lack of good solutions that are readily available. Use reliable data and reputable statistics to demonstrate a demand and unmet need for the product. Be sure that the logic is sound; do not draw conclusions through illogical assumptions about the market. Test out your assumptions on a skeptic to see if they are convincing enough to them. In this section you want to answer these questions:

(a) What is the problem that this technology or product will solve?
(b) Who are the potential customers?
(c) What is the size of the market?
(d) Why will customers want to buy this product (see Chapter 7 on Marketing).

Determining the size of the market can be done though credible sources. If your target market does not have industry figures available, use an alternate means to estimate the size of the market through population demographics, disease prevalence, or by combining of substitute product markets. Estimate the long-term growth prospects, and use independent market research for support. If there is partnering interest from reputable companies, cite these potential relationships, but do not overstate their enthusiasm. Unless a company has entered into some advanced and detailed discussions with partners of mutual interest, do not use initial meetings or phone calls as proof of interest.

5. Technology or Product Description

Remember the audience. Those reading the business plan are not researchers, scientists, or physicians and engineers. In this section be sure to:

(a) Briefly describe the concepts of the technology in easy to understand language that nontechnical individuals can read and understand – utilize graphics and visual aids to explain difficult concepts. Only explain the important concepts, not the details of everything.

(b) Focus on the results. Establish the credibility of the science and technology used to develop the product, but do not go overboard, this audience is interested in how much value the product or service provides to a potential customer.

(c) Explain how the technology or product solves the market need problem. Describe why the product is important for healthcare, patients, society, and why it is unique. Outline what the product can do that other products have been unable to do. Explain why and how customers will use the product and pay for it.

(d) Describe the most significant technical hurdles and how they will be overcome. Include development milestones and the process toward product development.

It is good to include a Gantt chart or a product development outline tied to estimated times of completion. Answer the question of scalability for the market intended.

6. Competition

Describe the competitive landscape and its impact on what the company is doing. It is important to be very thorough with market research on competition because you do not want a potential investor to tell you about an important competitor you should have known.

(a) Describe direct competitors and market substitute competitors (those with substitute products that are used by the target market).

(b) Outline how far the competition is from commercialization. If there are no current competitors in the market, explain why the company has a better chance of success then others.

(c) Describe the important differences of the product, not technically, but in relation to the value it provides to the market. Do not discuss differences that the market does not value. If the product does something great, but the customers do not put much value in that – do not talk about it.

Do not just say there are no competitors because this means that there is not an interest in the market – there are no competitors for counterfeit $3 dollar bills. Although a company may not have direct competitors, they will have ineffective substitutes for a specific target market, or they may have unsatisfactory substitute products that do not fulfill the market needs. What are the substitutes that your potential customers use? Does this substitute usage demonstrate there is a market need for a superior product? If there are direct competitors, possibly they are unsuccessful for one reason or another, explain why your product and company will be successful. Does your product or service have superior efficacy? If it is a diagnostic, does it have a greater accuracy, sensitivity, or ease-of-use? Know about companies with successful business models that are not directly competitive, use these examples as supportive evidence that your business model can also succeed.

7. Market Strategy

This is the plan for reaching the target market and how the company will execute this strategy. Refer to Chapter 7 for more information on market strategy development.

(a) Outline the expected time to commercialization and how the company plans to capture their market. Describe what makes you believe this can be done. Show the market research so investors can see the logic in the approach.

(b) Validate, with evidence, why the product will be purchased by its target customers. Describe how the product will be sold and how distribution will be handled. Outline any partnering or strategic alliances, and why they would be interested in partnering with your company.

(c) Describe how the company will deal with reimbursement issues and why this will work. Reimbursement is one of the most important and significant issues in the introduction of a new biotechnology product. Be sure to thoroughly research and understand how your product will be viewed by private insurance carriers.

(d) Show the company pricing strategy and why that price can be obtained for the product or service. Show who the target market customers are and how they make buying decisions. Show compelling reasons why they would make a decision to buy your product or service.

8. Financial Summary

In this section, describe what the financial projections are based upon and why they are achievable. Be aware of skepticism with revenue projections because anyone can plot sales that increase exponentially. Demonstrate a good understanding of the important drivers of revenue.

(a) Show revenue projections with gross margins, operating expenses, net income, cash flow, and a balance sheet for five years out. Do not go thorough detailed monthly financial projections because biotech investors know that this kind of detail is a figment of one's imagination. The only exception to this may be for diagnostic and medical device companies seeking later stage funding if they are already in the market.

(b) Include summary financials of the company in broad categories; these do not need to be broken down much. For instance, include: General and Administrative, Research & Development, Legal and Patents, Sales and Marketing (if relevant).

(c) Describe how much money the company intends to raise in this current round, and what are the uses of proceeds. Include the milestones this financing will allow the company to reach. Include the funding history of the company.

(d) Describe the number, and total amount of capital anticipated from all future rounds of funding, and when they are needed.

9. Intellectual Property

Include the company's patent portfolio, the patent firm's name, the stage of the patents (issued, CIP, PCT, International), and a Freedom-to-Operate assessment. If the FOI requires additional licenses from other patent holders, indicate this and describe where the company is at in obtaining these licenses.

10. Regulatory Information

Describe the regulatory pathway for the product and how the company plans to successfully complete the regulatory phase. Include any external factors that could improve or hinder the regulatory approval of the product. State what the company is doing to overcome or take advantage of these external factors. Describe any regulatory changes occurring in favor of product approval that may help reduce this risk.

11. Management Team and Background

This section is one of the most critical to raising money. Sophisticated fund managers will review this section first to see if the management team has the experience and credibility to make this venture successful. Many investors will plainly say they invest in people, not ideas because they understand that experienced and successful people can take mediocre ideas, and turn them in to successful products. Ideally, this section should show that the company has the people with the appropriate skill set to make this plan succeed.

(a) Describe the existing management team and their relevant experience in areas that are critical. Were they founders of previous companies? Describe how these individuals are capable of helping this endeavor succeed.

(b) Does the company have notable expertise in this industry based upon their management, Board, or Scientific Advisory Board and consultants?

(c) Describe the plans to hire key staff. Outline when the company intends to hire them and what expertise will be sought.

Be honest with the management team's strengths and limitations. Most start-up companies cannot afford to immediately hire the expertise needed to fill all management team positions to lead the company. The company may intend to hire a qualified consultant in a transition role until they can bring on a full-time person. If this is the case, state it in the business plan. Acknowledge which positions will be filled by consultants until seasoned individuals can be hired to fill those positions. Discuss key players and point out their accomplishments. If someone has been successful in negotiating licenses from major pharmaceutical companies, and the company is seeking alliances, be sure to point this out. If there is a medical device executive who previously grew sales at another company from $2 to $100 million dollars over five years, highlight this accomplishment.

12. Risks of the Business Model

There is no requirement for a section on Business Model Risks. Just be aware of the business risks and have plans to mitigate them because questions will be asked;

Venture Capital investors know most all these business risks. If there are risks to the business that always come up, help yourself by addressing them in the appropriate section. If the risk is a changing regulatory environment, address this by demonstrating that the company has either some credible inside information that supports this decision, or there is a back-up plan that is equally viable should these risks occur.

13. Return on Investment

Explain the company exit strategy and the timeframe anticipated for this to happen. Based on such an exit, estimate what the potential return on investment is for an investor. Most drug development and biologic companies focus on a public offering or acquisition by a major pharmaceutical company, whereas most medical device company exits are mergers or acquisitions.

There are various ways to estimate a return on investment for your investor. One is to first show comparable businesses that are either purchased or publicly traded, and what the market capitalization is for that company. On the basis of the company's current premoney valuation, assume a similar exit and then calculate a return on investment. Other ways may include calculations based on projected revenue. Be aware that projected revenues are viewed skeptically. This section must make a real investment case for a significant return even if the company falls short of all their revenue projections.

The business plan is used for the purpose of raising money. In the US, there will be other information necessary to include such as investment risks, which are required to be compliant with the Securities and Exchange Commission. This is where a good securities attorney will be of help to you.

The information described here may seem daunting for those who have never written a business plan. Most likely, help will be needed. The entrepreneur can hire a good consultant, or they may find help readily available through local government programs with assistance for technology start-ups. Some local governments may even have a commercialization center that facilitates the needs of technology start-ups, and they may have a service to help with the business plan. Start writing the plan yourself, and then seek professional help to finish or finalize it. The business plan should be a good representation of the company's capabilities and the teams' business acumen.

The Presentation

All That is needed to Raise Money is a Good PowerPoint Presentation

Although this statement sounds absurd – there is an element of truth in it. The business plan and Executive Summary is what secures a face-to-face meeting. I do not know anyone who invested in a biotechnology company after reading a business plan. Investors want to talk with the people behind the venture. If a Venture Capital group gives you the opportunity to make a presentation – congratulations! You have at least made it to a stage beyond many entrepreneurs. The presentation, its content, and the management teams' interaction with the investors is what will determine whether or not they take a closer look at the deal. Therefore, it is important to have a concise, balanced, and strategic presentation. The entrepreneur will get lots of practice because most investor presentations do not result in an investment. During the first five years of company development at a recent biotech start-up company, I gave over 100 presentations in order to various investors in order to close three financing rounds.

What is Contained in Your Presentation

A presentation will be similar in content to the business plan; however, it is important to be much more concise in order to capture the investor's attention. The entrepreneur should have at least two prepared presentations, one that is about 15 minutes and another version that is about 30–45 minutes long. They both should contain the information outlined below; the difference is just how much detail you provide. Later there will be a need to refine and produce specific presentations to specific audiences, such as those that will be made to strictly financial, marketing, or technical/medical representatives.

1. Company focus and history
2. The Problem that will be solved (why is it a BIG problem)
3. The Solution – The Product arising from the technology
4. The Technology, include a slide or two with supporting data
5. Intellectual Property protection and regulatory issues
6. The Market Opportunity and Competition, how reimbursement will be addressed
7. The Target Market and how it will be reached
8. The Financials and Proforma Estimates
9. The Timeframe for Development and Marketing
10. The Management Team's Experience, Board Members, Scientific Advisors
11. Summary Slide

Pointers on Giving a PowerPoint Presentation

- Limit a presentation to 15–20 slides depending on the time limit. Assume that it takes about one minute per slide. It is more difficult to be concise, so carefully select the most important points, in order to cover all the aspects that are listed above. Have back up slides for reference or questions. Do not bore investors into saying "no." Do not talk fast.
- Have a thought-provoking, attention-getting first slide. Investors, especially venture capitalists, sit through too many boring presentations. If boredom could kill, they would have already exhausted the lifespan of a dozen felines.

- Introduce each slide by saying why it is there – what is the point of the slide? This can be done by asking a rhetorical question as a lead-in. Do not just "data dump" the information from each slide. Investors can read the text without the presenter there.
- Use key words or phrases in slides as springboard talking points. Do not read the slides.
- Incorporate good visuals such as charts, graphs, pictures, technology images, etc., and use citations for credibility of information.
- Do not overuse text animation – less is more. It is fine to use simple builds for keeping the audience focused on the point trying to be made. Do not use sound animation for text.
- Make sure you have good contrast in slides so that it is easy to read from the back of the room.
- If the technology is complex or difficult to understand, it is helpful to have professional animation made to show how the technology works.
- Focus on content – just having pretty slides do not sell. Entrepreneurs must show they have the intellectual and strategic ability to lead the organization by addressing the important issues of the investors.
- Have a summary slide to review the key points of the talk so that the audience remembers these points.

Presentation Body Language

- Face the audience when speaking, do not turn your back to them and read the screen.
- Look at, and make eye contact with ALL the audience. Do not stare at your shoes or the computer.
- Do not use notes because you are expected to know this information. If you absolutely must use notes for some reason, glance at them for reference, then continue to look at the audience.
- Use your hands to point to the slides, or gesture as you talk about the slides, point and say "as you can see from this slide…" Do not talk with your hands in your pocket.
- Speak up when presenting. Everyone needs to hear you.
- Do not use the word "umm"or "uhh". Rather, use silence and pauses. Practice giving presentations in front of a video camera and see if you impress yourself.
- Do not speak in a dull monotone voice. Put some excitement and enthusiasm into what you are saying, but not hype.
- Stand on both feet and do not nervously weave or rock – it is distracting. Walk or change positions.
- Be yourself and act natural, work on being comfortable with the investors you are talking with. Hopefully, you will be working with them in the future.

Other Presentation Points

- Know your material! Do not worry about saying the right phrases or sentences. DO NOT memorize, because it will *sound* like you have memorized the information. Talk to the audience as you are telling them the story.
- Dress professionally. You would be surprised how much investors pay attention to this.
- State why you are there: "I am here to raise $5 million," "...to find a partnership," "...to introduce our company for future fund raising."
- Practice in front of friends and family, get comfortable with your slides and your material.
- Prepare and anticipate technical glitches. Bring a back-up of your slides on a jump drive, and bring a few handouts just in case you do not get to use slides.
- Most of your presentations will have a time limit, and most of the alloted time should be spent in Q&A. Plan your time accordingly.
- When presenting at a first meeting with a venture capital group, *do not* leave without asking for a date when you can return for a follow-up with additional information.

Building a Web Presence: Putting Your Best Foot Forward

The first impression an investor will get of an organization and its level of sophistication is determined by its web presence. A company website is the "brick and mortar" of the twenty-first century. A high-technology company cannot be taken seriously without a quality web presence. Thanks to the internet, a start-up company can have the look-and-feel of a successful, sophisticated, multinational organization, and no one would be the wiser (until they call the main number and the CEO answers the phone). The point is that without spending a lot of money, a start-up company can get a jump start on its positioning by its web presence. Invest time and careful thought into the design of a website and its message to the target audience. I am not talking about spending your annual budget on a new website, although that can easily be done. Find a good web designer and creative artist to help at a reasonable cost. With all the new self-help software, the entrepreneur can even do this themselves, although I would not recommend this unless they did it in another life. The entrepreneur's time is needed for other things, and generally professionals can do this better and faster.

When building a website make it open-ended and leave room for expansion and improvement. Avoid the use of "under construction" dead end web pages. Put the page in later, and do not tell the world that the company has not completed something – it gives the impression that you are an unfinished company. When a company has an impressive, good quality website, the audience assumes that they are a high quality company. Remember the saying "you never get a second chance to make a first impression." Make the first one good, by having a steller web presence.

Final Comments About Funding

A tremendous amount of time and effort are put forth to secure funding at the right time from the right financing partner. Because of the enormous expenditure of time and effort, sometimes funding events subtly become *the* goal of the company. Never forget that a funding event is a means to an end, and not an end in itself. The company goals are product development, market development, and business development milestones that increase the value and reduce the risk of the company. It is wonderful to celebrate funding events when they occur, but funding is just a baseline, albeit critical, obligatory necessity, which allows the organization to have an opportunity to achieve its *real* goals. As humans, we need oxygen to live, if we do not continually get oxygen to our brain cells, within two minutes the death process begins. It would be absurd to hear someone say that their life goal was "to breathe oxygen." Keep your focus on your real goals, and make consistent progress so you will not lack for future funding.

Sometimes new founders and CEOs have a sense that when they compete their fundraising – the hard work is done. This sentiment may account for complacency and false security that may occur after funding. A company's intensity may even downshift after they achieve funding, and there can be a tendency to coast until they get closer to needing the next round of financing. Obtaining funding just means the hard work can begin in earnest. The leader must always maintain the organization's intensity toward reaching the next value-increasing and risk-mitigating milestone, *if* they hope to have a successful biotechnology company.

Chapter 10
Corporate Culture and Core Values in a Biotech Company

Companies are unique organisms with their own personalities. This does not mean they can become life-form monstrosities that eat Philadelphia – it just means they actually have unique personal characteristics that are specific to each organization. Typically these combined characteristics are referred to as a company culture. A company's culture is so tangible that individuals can sense a real difference between one company's employees and those of another organization. For instance, one particular company may have a culture of innovation, excitement and expectancy, motivating employees to ignore the traditional work hours, and do anything to get the job done no matter what time of day. Conversely, another organization may have a minimalist culture with clock-watching employees doing only what is required if, and only if, it is written in their job description.

Why is a Company Culture Important?

A corporate culture can be a strength or a weakness. Since every company will encounter problems and challenges (biotechnology companies have more than their fair share of them), their culture is a good indicator of their ability to overcome crisis. Ortho McNeil, a Johnson and Johnson company, is well known for their response to the "Tylenol scare" of 1982, where tainted Tylenol Extra Strength capsules were found to contain cyanide that resulted in the death of seven individuals. It was quickly discovered that these capsules were tampered with *after* they had left the manufacturer, and it was not the company's fault; however, Ortho McNeil made the decision to immediately recall 31 million bottles of Tylenol at a cost of over $100 million dollars. To further protect the public, they advertised nationally, not to consume any products containing Tylenol. Ortho McNeil employees responded, based on their corporate culture of putting first the well-being and needs of the people they serve. This situation is analyzed in business schools around the world as an example of how to successfully respond to a crisis. Most interesting is that crisis management was neither really "taught" to the Ortho McNeil employees, nor was this response

C.D. Shimasaki, *The Business of Bioscience: What Goes into Making a Biotechnology Product*, DOI 10.1007/978-1-4419-0064-7_10,
© 2009 American Association of Pharmaceutical Scientists

rehearsed. Their organization's culture encouraged and lived the core values established by their founder, which guided their day-to-day decision-making. Their company culture and core values only became visible to the public as a result of this crisis.

Corporate cultures are based on foundational values. In order to have stability, an organization needs foundational values that are clear and unwavering. A company without core values is equivalent to a ship without a compass. Yes – it is possible for companies to operate without them, and work can and will get done. The issue is not whether things get done, but rather *how* and *what* things get done. In this chapter, we discuss the importance of company culture and core values, and their significance in helping to build a successful company.

Development of a Company Culture

For the biotech start-up company, developing a corporate culture may not seem like a high priority compared to the need of finding money, hiring a team, and quickly making product development progress. However, it is critical to pay attention to the development of a culture because at some point the company reaches a critical mass, and like concrete, the culture of the organization becomes set. One can temporarily ignore the development of a culture without much consequence, but at some juncture this will impact the company's ability to get work done effectively. For those just beginning a company, make mental notes of these things. If the company is fully developed and it is evident that the culture is counter-productive to progress, now it is time to do something. One thing is certain – there *will* be a company culture – the question is whether it is by design, or default. Choose to purposefully build a culture that adds strength, rather than is divisive by default.

The culture of a company can help or hinder it from successfully making changes. All companies go through transitions and growth phases (as discussed in Chapter 14), and everything does not always happen exactly as planned. Changes and adjustments are required for a company to grow – just talk to employees who have been with a start-up company for several years. I recall one early employee talking about how five years previously, they did all the accounting, payroll, market assessment, public relations, and administrative work – and *missed* that. What they really missed was the smallness, having multiple responsibilities, and the daily dynamics of variety. As companies grow, they subtly change the way they operate, but people who share the same core values, adjust to these changes while maintaining the same core values.

Most start-up biotechnology companies begin with an entrepreneurial culture. There is excitement and anticipation within the company about new things. Employees have expectations of professional opportunity, personal rewards, and aspirations of contributing to a greater good. These are expectations of the future. Each organization should develop a culture suitable for their business model and particular industry. A one-size-fits-all corporate culture does not work. A company that operates in a highly-ordered industry such as accounting, is not best suited to have a culture rooted in creativity, but rather on constancy – yet there are other qualities that support core values common to both.

Individuals Define the Culture and Core Values

An organization's strength is the sum total of the individual strengths. Individuals make decisions based on knowledge (which is generally what we look for when we interview and hire an individual), and their core values (something companies usually do not look for during an interview). An individual with the core value of "mutual respect for others" will make a different decision about leaving an unfinished job for another employee, whereas an individual with the core value of "acceptance of responsibility for one's actions" will readily admit a mistake rather than cover it up or blame someone else. These instances may seem trivial by themselves, but add these up amongst 10, 100, or 1,000 employees and this becomes the culture of an organization. We all have heard similar statements such as "I don't like working with that organization, because of their people" or "I really like their products, but I just don't trust the management." A fundamental error of business is to only focus on products, without consideration to the people who produce the products.

To build an organization with a desired culture, the company needs a hiring process that includes careful selection of new members added to the team. A rapidly growing organization will hire the equivalent of its entire company many times over, until it reaches a critical mass. During growth phases, a company can either ensure that their culture is maintained, or they can allow it to transition to the least common denominator of new employees. Hiring individuals that do not share company core values results in some employees driven by motivations not shared by others; these individuals will come to different conclusions based on the same information. Differing core values cause internal conflicts, which would have been disastrous for Johnson and Johnson during the Tylenol cyanide crisis. Recognize, that the hiring process is also a gating mechanism which the rapidly growing organization should use to ensure the maintenance of its culture. However, the leaders of the company must first identify what constitutes a "fit" employee. This aspect is discussed in Chapter 11.

Examples of Core Values

Below are the examples of core values and guiding principles embraced by one biotech company. These are not the only ones, nor are they necessarily the best ones. The key is that an organization has core values that its employees espouse and live. The selection of core values directs the type of corporate culture a company will produce. Core values can be thought of as the seeds of an organization's future. Someone planting corn seeds would not expect to get watermelons. The selection of a particular set of core values later bears the fruit of those values.

Company Core Values

- **We are Team Players:** We help each other grow within and across departmental functions to be part of a greater purpose; it is not just a job.

- **We Show Mutual Respect to Everyone:** We value each other and it is exemplified by our behavior. We listen to each other, using tact when appropriate and giving reasons why we do the things we do.

- **We Value Honesty and Integrity:** We say what we mean and we do what we say, whether it is convenient or not.

- **We are Accountable and Responsible:** We take ownership for our actions, both right and wrong.

- **We have Open Communications:** Across all levels and functions of the company for internal and external information, both good and bad news.

- **We are Focused:** We have defined plans and ideas that are prioritized to reach our goals. We have the discipline to say "no" when appropriate.

- **We are Empowered:** We perform our tasks with creativity to reach our goals while developing new solutions to problems. We understand the value we contribute to the overall goal of the organization.

A set of core values must be defined and agreed upon by all. Core values are even more important to an organization's future than a mission statement. Mission statements are standard convention, and it is a rare organization that does not have one. However, core values cannot be treated like many company mission statements. Too often, mission statements become like "employee of the month" awards – it sounded like a good idea for a while, but later everyone lost interest. I chuckled to myself one day when I entered a company lobby and saw a dusty old plaque with a beautifully inscribed company mission statement sitting in a corner where few could see it. Next to this relic, but prominently displayed, was their "employee of the month" plaque. The irony was their last employee of the month was awarded in February 1999 – my visit was in December 2006. If the company ascribes worth to their core values, the employees will also. One biotech company wanted their core values to be so integrated into their team, that they provided a personal day off for those that would memorize them and recite them during a set period of time each year. This event let the employees know that their core values were not an afterthought to the company but an integral part of their workplace.

Guiding Principles Based On Core Values

Core values translate into various guiding principles upon which individuals in that organization make decisions and operate. When management holds themselves and each other to a set of core values, the organization operates in a consistent manner, and there is certainty as to how problems and issues will be handled. A set of guiding principles can be defined, or they may just be understood. Below are some examples of guiding principles based upon various core values:

- We are a relationship-oriented company. We incorporate into all our business, market, and science endeavors a relationship orientation with our "partners" to achieve common goals that benefit each other. We will seek relationships with partners that share common, or mutually beneficial goals.

- We differ from our competitors by incorporating innovation into all of our products. In addition, the products that we develop will not be an end in themselves but tools to improve healthcare in the broadest sense.

- When our products are launched, we work to improve outcome measurements such that the true long-term benefit of our products will be understood.

- Our profitable differences will be seen in our product innovation value discipline and not in the commoditization of products. Our product positioning will be "best in class" where there are competitors. This will be accomplished through our products themselves, or through the collateral services and programs provided with our product, which creates more value to the customer than the product alone.

- We highly value our partners and patients, and are loyal to them and their needs. Our partners are our physicians, center employees, vendors, and suppliers. Our patients are our mission.

- We are a diverse group of individuals, with differing backgrounds who share the same core values. We encourage each other, and also remind each other if we do not live up to our own values.

Core Values Begin With the Leader

Building great organizations require great people, and great people require great leaders. Core values start with the organization's leader. They define the core values that the organization espouses, such as the Credo[1] developed by Robert Wood Johnson in 1943 for the Johnson and Johnson Company. Core Values are the belief

---

[1] Johnson and Johnson Credo Values, http://www.jnj.com/connect/about-jnj/jnj-credo/

system and set of behaviors, individuals hold that are nonnegotiable. Core values represent the real person. It is who they are when they think no one is watching.

Dr. Samuel Waksal was a budding life science entrepreneur. He had knowledge, experience, and the right combination of technical background and business savvy. He was fortunate enough to work in a well-respected laboratory and knew when he saw something of potentially great value. He licensed a product opportunity, and was able to convince financial institutions of its value, and he and the management were able to build a seemingly successful biotech company. The company went public in 1991. However, during the growth of the organization the core values of its leader remained. These became evident to the public when Dr. Waksal was notified by the FDA in 2001 about their product not being approvable. This started the ensuing insider trading by him, his family and Martha Stewart. Later, Samuel Waksal was indicted for obstruction of justice, bank fraud, perjury, and insider trading. Fear can cause individuals to make poor decisions, but strong core values help keep them from making unwise decisions – or after making a bad decision, it lets them admit a mistake. Strong core values permit a person to sleep at night, accept circumstances, and respond in an appropriate manner. It is like having a compass, when lost – a person may not know exactly where they are – but they know the right direction to take.

Strong core values do not automatically ensure business success; however, they provide the *highest likelihood* of success; and when successful, they ensure it is sustainable. When desiring to build an organization of lasting value, its leader's core values cannot be ignored. No group or team can be effective over the long-term without sharing similar core values. Strong core values are beneficial beyond business and into family life and other relationships. A good book I recommend and have given out to all my employees, contains simple truths written by an ex-school teacher, and can help identify core values you may want for your organization. The book is titled "Life's Greatest Lessons: 20 Things that Matter."[2]

How to go About Improving a Company Culture?

When corporate cultures are in disrepair, many good efforts are made to change the culture by implementing new programs, adopting a mission statement, or by bringing in consultants to help change the organization's behavior. Sometimes this improves the situation for a period of time, or it may just superficially change behavior. However, like a physician who treats food allergy symptoms but does not identify and remove the offending substance, most often the results are temporary. There is always room to improve the culture of any organization. There are many things that can be done to improve a culture, but if the management and leaders of the company do not think it is important, these will not last. Improving corporate culture cannot be delegated to HR – it must come from its leaders. In order to improve a company's culture there must

---

[2] Hal Urban 2003. 4th Edition. "Life's Greatest Lessons: 20 Things that Matter". Fireside Book, New York, pp. 165.

first be a driving desire for lasting change from its leaders. There are many books on the subject of the characteristics of good leadership, so we will not spend time on this here – just note that people follow what leaders do, and not what they say.

Communication in a Biotech Company

Communication is the lifeblood of all good companies, but communication becomes even more vital to a rapidly growing biotech company. In a dynamic company, direction and initiatives can change rapidly. Without frequent and effective communications, problems arise that impact the effectiveness and efficiency of scientific and business progress. Companies need to have frequent and regular forms of communication, and these forms must change and adapt as the organization grows. A company of 24 employees has different communication channels than a company of 4 or 400. Good communication is supported through multiple forums to exchange ideas and share views. The obvious ones are frequent meetings, e-mails, telephone messages, and impromptu exchanges. It goes without saying that companies can also overdo meetings that either have no value, or are attended by individuals who do not benefit from them. Keep formal meetings to the minimum necessary to accomplish the communication goal. Larger companies have the benefit of additional communication vehicles such as intranet, blogs, and electronic forums. Balance the forum with the effectiveness of the communication based on the size of the organization. For instance, using an intranet to communicate when the company has only 10 employees will seem impersonal, though it might be a great way to have all the information in one location. Good communication is an important company core value, and other core values are made evident through communication. Without communication, there is only inference as to a company's core values.

Core Values Apply to Partnerships Even Suppliers and Service Providers

Support of the company's core values must extend to all aspects of the business; it should guide choices in partnerships, alliances, and even the selection of critical suppliers and service providers. The selection of a critical supplier or service provider should be approached in a manner similar to the hiring of a senior or executive staff member of the organization. The same care, scrutiny, and alignment with common goals should be examined prior to proceeding with a relationship. The 80/20 rule applies here, in that 80% of one's time will be spent on 20% of the worst relationships. As the organization begins to grow and makes progress, a company will become dependent on many external relationships. The last thing a rapidly growing biotech company needs is to experience break-downs in relationships with critical partners – unfortunately, this is when problems most likely occur. The best time to evaluate a relationship is before it is consummated.

Companies spend too much time dealing with ineffective relationships, which are manifestations of inconsistencies in core values between the two parties. These inadequate relationships divert management time away from building the business and reaching company goals. All relationships will at some point encounter difficulties, but these difficulties should be opportunities to improve a process, a communication channel, or improve the working relationship, rather than a substandard way of doing business. When substandard behavior is accepted as normal, these relationships become exhausting and burdensome (Strategic partnerships and alliances are discussed in more detail in Chapter 12). As a result, most team members will go out of their way to avoid working with those in this type of relationship.

It is always a good idea to seek multiple suppliers and service providers for any critical need. Do not make decisions on pricing alone; evaluate their communications, strategic directions, corporate values, and interest in a long-term partnership rather than solely as a business transaction. Ask yourself the following question, "Do I think I can depend upon this group to come through for me in difficult situations?" If the answer is "no," or some variation of uncertainty, it would be best to identify alternatives before making a committment. Good suppliers and service providers are willing to work through issues, and concede short-term financial gain if you have a partnership orientation and an interest in a long-term relationship benefiting all. Conversely, those that only seek a single transaction will not show flexibility in working with your organization, and they should be avoided.

Be aware that the quality of service, interactions, and communications with any partner or supplier tends not to improve beyond the early interactions. One may see improvements for a season, but in the long run these usually revert back to a standard of business practice – good or bad. Sometimes a company is forced to stay with a supplier or service provider for reasons that are not ideal. In these situations, the company needs to make it clear how their organization views this relationship, and what the other must do if the relationship is to continue. Always communicate a long-term partnership goal, and it will reduce the time spent on any one particular issue, for it allows individuals to find reasonable solutions together. Always let suppliers and service providers know where they stand, and give them feedback and recognition for their effort.

A Good Corporate Culture is a Strength That Cannot be Easily Copied

Do not overlook corporate culture as a strategic asset of the company. Any business model, organizational process, or business strategy can be copied by a competitor. However, it is almost impossible to copy a corporate culture. From a business perspective, it makes sense to develop a corporate culture that provides the company with a strategic advantage. Cultures are like the DNA of a company – it is hardwired into their system and can only be sustained if it is genuine.

Chapter 11
Hiring a Biotech Dream Team

It is impossible to begin a biotech company and build it successfully without the help of a myriad of individuals and external resources. Seasoned entrepreneurs and executives all reach a point in their career where their success is no longer determined by what they can do themselves, but by what they accomplish with, and through the work of others. The previous chapters may seem to have been directed toward a "single" biotech entrepreneur – in reality, successful biotechnology companies are usually established by a growing team, and not by one individual.

Building a successful biotech company requires an ever expanding team. Therefore, it is important to find the right mix of individuals with complementary expertise, having the leadership characteristics necessary to establish and grow an organization toward a common goal. The careful selection of team members is important because (1) they provide the expertise essential to reach these goals; (2) their expertise is critical to securing investment capital. Choose a team wisely, because the credentials and experience of the team are viewed as indicators of future success by venture capital.

For the Start-Up: Virtual is Vital

I am a strong proponent of a "virtual company" during the start-up phase of a company. A virtual company simply means that the company does not own or lease a physical building, and it does not have full-time employees – yet the company still conducts all functions as if they did. The company may be borrowing equipment or bartering for time on equipment, and the facilities may be the academic laboratories of the scientists, and a post office box to receive business mail. In addition, each member of the team usually carries multiple responsibilities, and this may not be their day job. Virtual companies usually parse out various product development testing and market research work to outside entities.

C.D. Shimasaki, *The Business of Bioscience: What Goes into Making a Biotechnology Product*, DOI 10.1007/978-1-4419-0064-7_11,
© 2009 American Association of Pharmaceutical Scientists

At the start-up stage of a company there is little money to hire anyone, let alone pay the entrepreneur a salary. However, progress must still be made to gain interest from investors. In the beginning, the company may consist of cofounders or Scientific Advisory Members and Board members. Others may be working under contracts or temporary agreements. One non-cash way to secure help for critical work is to consider incentives such as stock, stock options, or restricted stock as compensation for professional services as discussed in Chapter 5. When doing this, be sure to discuss all employment and compensation plans with your attorney, as they will give guidance and help drafting the necessary documents to prevent problems with ownership and issues related to intellectual property protection.

Being a virtual company provides time to raise Start-up, Seed Capital or Series A round financing, and also allows time for the company to improve its valuation as development progress is made. Virtual companies conserve large amounts of cash that should be used to advance the technology, rather than support a large overhead that limits their existence. Once a virtual team is assembled, finalize the business plan, and work to secure financing as discussed in Chapter 9. Making good progress and completing proof-of-concept work improves the chances for an early capital round such as a Series A Preferred investment.

It is always important to retain the scientific team that developed the technology in some capacity. If this is not possible, at least retain any critical member, and appoint them to one of several positions as described in Chapter 2. At one start-up company, we preserved the core technical team by leasing these individuals from the research institution where they remained as employees. The institution allowed us to defer some costs until we secured capital, and they received some equity in exchange for this consideration. Having a good relationship with the licensing institution can provide a company with some creative options for low cost development work.

Renting and Leasing Space

At some point, a company will indeed need dedicated laboratory and office space if they hope to effectively make significant product development progress, but stay as virtual as possible, for as long as possible. When making the move to dedicated space, check out the availability of any technology incubator space. Some incubators are subsidized, which allows a company to enter into office and laboratory space at reduced rates that later escalate over the time period of their stay. Incubators may also have some support services or shared equipment opportunities. If your company is located near a biotech hub, there may be opportunities to sublease space from similar technology companies in a research park. Be sure to carefully anticipate your space needs now and in the near future. Balance these plans against the capital the company has currently, and the length of time required to reach significant development milestones. Renting biotechnology space is not cheap. Depending on the location of your company and the mix of laboratory to office space, it may cost between \$30 and \$75/sq ft. per year. Be sure to know the impact of any long-term lease and how these obligations impact the overall burn rate of the company. Be aware of the time it takes to raise another round of capital.

Who do You Hire First?

Once a significant infusion of capital is secured, the company will need to hire staff. Be aware that staffing quickly consumes the greatest allocation of funds in a young biotech company's budget. If money was not a limitation, start-up companies could hire an experienced executive to lead every function available. However, it is rare that a start-up can immediately hire experienced Vice Presidents or Corporate Officers for every discipline. Although the company may need immediate help in all areas, senior level personnel is not necessary for all functions. It is essential to find the right mix of leadership and strategy contributors, along with task-orientated individuals. Consider the type of staff to bring on, at what level to bring them on, and whether full-time, part-time, or contract will suffice. Because of the enormity of tasks to accomplish, the biotech entrepreneur may begin to hire based on availability rather than on need – this is not a good idea. Success is inextricably impacted by the early individuals hired.

A good question to ask is "who do I hire first?" To help, I have listed nine functional areas a biotech company will likely need help in sometime during company development. These functions are not just individuals, but specific disciplines – although at some point, they will be filled by an individual or groups of individuals. These nine functions are subdivided into three groups we will call an "urgency factor." The purpose of prioritization is to solidify hiring needs and determine when to bring on full-time individuals and at what level. Although every function is essential, they all do not require the immediate attention of a full-time person. Whether the entrepreneur(s) realizes it or not, in the beginning they may be performing all of these functions themselves. Depending upon the portion of the industry the company is working in, there may be other functions needed, if so, they can be added and inserted into this list (Table 1).

Table 1 Categories of expertise

I	Business leadership
	Scientific/Technical
	Legal
	Market development
II	Medical/Clinical
	Finance
	Regulatory
III	Personnel
	Operations
	Business Development

Category I Functions

For the majority of biotechnology companies, Category I functions are urgently needed and consume large allocations of time and money, requiring a good portion of their budget.

The *Scientific/Technical* expertise is an obvious critical need for a biotech company, so hire the best and most senior level because the company value is closely tied to its scientific credentials. Many times this function is led by a Chief Scientific Officer or Vice President of Research. Additional senior level help may also come from a Scientific Advisory Board. *Legal functions* are urgently needed to properly establish the corporation, issue stock, draw up contracts, and carry out securities and patents matters; however, this function should always be outsourced to the best and most competent legal specialists. *Market Development* support is not the same as Marketing support. Market development refers to the expertise needed in defining a market strategy, determining a target market and assisting with primary market research into areas such as market need, reimbursement issues, adoption issues, pharmacoeconomic issues, and pricing (refer to Chapter 7). Young biotech companies will not be hiring a full-time market development employee very early on but they do need help in this functional expertise right away. The *Business Leadership* function is usually filled by the CEO or entrepreneur. If the entrepreneur does not have the business leadership capabilities or cannot grow to fill them, a seasoned CEO will need to be hired later as the company raises larger follow-on rounds of capital; sometimes this will be a condition of financing.

Category II Functions

These functions are also important, but during early stage development these should be filled by outsourcing arrangements or by occasional help from qualified individuals in the field. During the early stages, a company should get sufficient *Medical/ Clinical* advice and help from their Scientific Advisory Board members. This can be supplemented by additional contract professional help. For companies developing a therapeutic or biologic, as they move into later stages of product development they will then consider hiring internal support for this function, and usually this will be the most senior position, such as a Chief Medical Officer. *Finance* help at the formative stages can include the help of a part-time bookkeeper and occasional senior financial advice from a professional. As the company grows they can convert the bookkeeper to full-time (especially if they have grants) while outsourcing the CFO functions. At some point, the company will need an external auditor, at the early stages it is unlikely they can afford a Big 4 auditing firm; alternatively, retain a good mid-tier firm. As the company moves closer to major financing, and certainly a couple years prior to an exit, hire a top-tier accounting firm. Also, when the company gets closer to an exit such as an IPO they will need a capable CFO. *Regulatory* assistance becomes more critical the longer a company is in development, but early on they should seek the advice of experts on a consulting basis. As the company gets closer to any type of clinical studies, they need to hire the best regulatory experts they can find, but it may, or may not require a full-time position. This decision will depend on their product and the expertise needed. During the final stages of clinical trials there will be enough full-time work to support a senior regulatory affairs person. Always seek the most experienced regulatory expertise that is directly applicable to the product and the disease field in which you are working.

Category III Functions

Personnel, Operations and **Business Development** functions can usually be managed by the entrepreneur and core team during the early stages. Until there is sufficient capital and long-term prospects for the business, these positions do not usually become full-time until much later in the company development. Personnel, payroll, and benefit needs can be outsourced to a Professional Employer Organization (PEO, discussed later).

General Recommendations

These functions and categories can vary depending upon the business model, so do not take this as a prescriptive approach for the business, but use this as a guide. The key is to have a priority schedule for staffing so as not to fall prey to hiring the first available individual without considering the most urgent need of the company. Once the functional needs are identified and prioritized, then assess the seniority level and experience to hire. There are no hard and fast rules for this, and it is difficult to generalize but here are some guidelines:

- If the function is absolutely critical and it impacts any criteria for funding, hire the most senior level person with the most experience and best credentials possible.
- If the function is one that can be filled by many qualified individuals, hire junior level individuals first and give them an opportunity to grow with the company.
- If a function is one that is extremely difficult to fill or one that is so specialized that there are few candidates, consider the most senior level reasonable. Recruit them at a level that will attract them to the organization.

When hiring, consider also the future reporting relationships for each function and communicate these plans when employees are hired. Does the hire know that the company intends to bring on a more senior level person later? Let individuals know whether they will have someone reporting to them, or if the company will be hiring someone they will report to later. Do not lose key people because of mismanaged, or unfulfilled expectations.

These categories and urgency factors are simply a way to prioritized the hiring of staff, and to identify the various functional needs of a fast growing company. We have not attempted to address the total number of individuals nor the number of support individuals reporting to the first hire within function. This number would be dependent upon the organization and their particular needs of product development. Within each of these functions there will be subdivisions of related functions particularly in the Scientific/Technical function. The important point to remember is to prioritize hiring needs.

A young company can minimize future financial shortfalls by carefully managing the number and mix of full-time, part-time, temps, interns, and consultants necessary to effectively accomplish their goals. This allows flexibility and the ability to rapidly grow without taking on undue financial burdens or worse, having to later lay-off staff because of insufficient funding.

Employee Benefits Challenges

During the formative stages, a lean biotech company will wrestle with the challange of providing competitive employee benefits which are needed to attract and retain full-time employees. It is difficult to offer competitive benefits as a company with only a few employees. Larger companies can access great benefits packages and can afford to compensate their employees in a way that small companies find extremely challenging. However, a smaller start-up company can usually be more generous with their equity compensation and their assignment of responsibility to key hires. A good way to both manage and access a variety of employee benefits is to find a Professional Employer Organization (PEO). These organizations provide low cost group health insurance and give access to similiar competitive benefits that large companies enjoy. Most PEOs also provide employment law support, hiring and firing assistance, employment law interpretation, and can help implement an annual review process. These organizations also provide the virtual human resources functions, including payroll and related functions. Generally, they offer these services for a small fee structure based on a percentage of the payroll. There are several good PEOs with varying features and benefits, check them out and see if they may be right for your organization. One national company we used at two start-ups is Accord Human Resources, and they worked very well for us growing from 4 employees to 35 employees.

Key Employee Compensation

There are attractive and enticing advantages to working in a smaller company versus a larger corporation. These benefits include greater autonomy, more responsibility, the challenge of doing something new and exciting, and equity ownership. Although a start-up or small to medium size biotech company may offer more equity to their key hires, they must still compensate adequately to attract and retain employees. Compensation varies with the geographic location of the company and the competition for talent. Compensation for key employee may be higher in San Francisco and Boston than it is in St. Louis and Denver. Likewise, this would be true for cities such as London, Tokyo, and Dusseldorf compared to surrounding cities in their countries.

The best way to assess competitive salaries is to first compare compensation for similar jobs in your surrounding area, then add a factor to account for the critical nature of the position in order to increase your ability to attract the best candidates. We will not go into detail here about the specifics of compensation because there are many good salary surveys for various positions in the biotech industry broken down by country and locale. A good survey for biotech salaries in the US is the annual Compensation and Entrepreneurship Report in Life Sciences, which is a study produced by professionals at WilmerHale, Ernst & Young, and J. Robert Scott and Professor Noam Wasserman of Harvard Business School. This annual report gives data compiled from almost 1,000 biotech executives pertaining to key executive salaries, stock options for both founders and nonfounders, and bonuses broken down by company size and funding round. A copy of this report can be obtained for a fee at http://www.compstudy.com.

Diversity is Essential

This is a conundrum – a company needs individuals with vastly different skills who each speak a different jargon, yet they need unity to work together toward a common goal. Though difficult, finding and hiring the most talented and functionally diverse individuals is not the hard part. The challenge is bringing these individuals together and helping them work through differences, while moving rapidly toward a critical goal.

At a previous biotechnology company, we were focused on viral disease diagnostics and therapeutics with multiple projects all targeting viral enzymes. To simultaneously move all of these projects forward, we created five functional department disciplines to simultaneously advance each product in the pipeline. These departments included: Organic Chemistry, Protein Biochemistry, Virology, Cell Culture, and Enzymology. Each Department Head was expertly trained in their field, but some came from academic backgrounds with little or no industry training. In an academic setting, principle investigators are rarely forced into collaboration with others, rather they seek collaborations of their choosing, with other researchers having similar or complementary research interests. However, in the biotech industry all disciplines usually work on the same project, with each having different responsibilities for that same project. An interesting situation evolved where certain Department Heads rarely communicated with other Department Heads, even though they had critical interfaces for product development such as the synthesis of an enzyme inhibitor, and the subsequent testing of enzymatic activity and inhibition. Often the reason given was that they did not understand what the other did, nor could they understand their results. My job became interpreter, translator, and negotiator.

Diversity is critical to the success of any biotechnology company because specialized expertise is required for every facet of product development. But with diversity comes viewpoint differences, differing preferences, and challenges. In spite of these differences, all company functions must be *interdependent*. Without functional interdependence all corporate progress would cease. Such would be the fate of a young biotech company without interdependency among functions. Just as each organ of the human body is diverse with highly specialized functions, each company function must be interdependent with the other in order to thrive.

A New Hiring Approach

In addition to interdependence, a company must recognize if there is a "fit" within each individual they hire. When deciding to hire a full-time position, utilize a hiring process that identifies the best fit individual. All employers know the importance of hiring individuals with the most relevant experience for a position. However, other characteristics that are rarely considered are equally important. When interviewing candidates, at least three aspects should be examined for fitness: Relevant Experience, Ability-to-Execute, and Shared Core Values. The Venn diagram below depicts these characteristics as distinct but intersecting, showing their interdependence. The opacity indicates the ability to observe these characteristics during a typical interview process (Fig. 1).

Relevant Experience

All companies examine relevant experience when considering an individual for hire. Without question, relevant experience is essential to the position, and it is the easiest attribute to identify and evaluate. This characteristic is important, but unfortunately it is many times the only factor examined. Relevant experience alone should not make anyone "a qualified candidate" for the job, but it should be one of three required characteristics examined before considering them as a qualified candidate. Relevant experience qualifications include depth of experience, academic credentials, and work experience. These are well understood, therefore we will not elaborate on this characteristic much further.

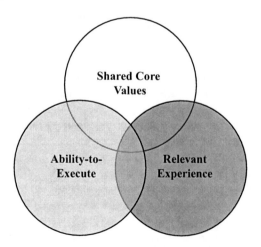

Fig. 1 Interrelationship of characteristics in a fit employee

Ability-to-Execute

The ability-to-execute is a characteristic that represents "doing" rather than talking or thinking. Traits such as "finishing a job started," "always coming through," are character qualities of an Ability to Execute. Suitable candidates should have relevant experience *and* possess the ability-to-execute. This combination ensures that a person can think of ideas, form a strategy, and execute a plan. People who lack these attributes, usually talk the language but cannot complete or finish even the simplest of tasks. The ability-to-execute (or inability) is typically discovered several months after hiring an individual. People without this attribute are individuals we think are getting things done because they can talk the talk, but upon examination they usually come up empty-handed. Such an individual becomes a time-consumer for management because the ability-to-execute cannot be taught on the job, and management cannot help them complete every single task. These individuals divert attention, and require constant monitoring and maintenance for fear that their project may be in jeopardy.

Although the ability-to-execute characteristic is difficult to identify, a good interview process helps uncover individuals with this attribute. Having the ability-to-execute and relevant experience is an important combination that ensures a person is able to deliver what is needed. However, having these two characteristics, still should not make any individual a qualified candidate for employment.

Shared Core Values

A person with relevant experience, the ability-to-execute *and* shared core values make a qualified candidate. Without shared core values, two people cannot effectively work together for a long period of time, nor will they work to their full potential. Core values are as diverse as people. They include qualities such as "a good work ethic," "responsibility to get the job done," "honesty in all situations," "reliability," and "loyalty to the company." Everyone has core values – even mafia or gang members, although their core values may be "get what you want at any cost," "deception is justified," and "having a conscience is a sign of weakness." Oddly enough, those negative core values tend to unite these groups, and a member without the same core values weakens the "team" from accomplishing its objectives. No team can be effective without sharing some of the same core values. The most effective teams are comprised of individuals who share a maximum of core values, and they influence other members to accept any additional values that are important to the rest of the team.

Core Values can be thought of as our personal "well of conscience" from which we draw when we face difficult situations, crisis, or problems, especially when we do not immediately know what to do. Having shared core values is the reason the Ortho McNeil team responded so quickly and in unity to the Tylenol tampering described in Chapter 10. To have an effective team, individuals must draw from similar wells. When problems are encountered, all responses need to be consistent with the company values. The most successful team leaders are the ones with the ability to recognize character qualities that match their team's shared core values. Good companies learn to identify shared core values in the people they hire.

The Hiring Process

A company's hiring process should be the best practice for evaluating the three characteristics discussed. The interview process should be structured in a way that allows the company to learn about the person, rather than just hear the right answers to specific questions. For instance, it is obvious that a person cannot assess the interpersonal or verbal communication skills of an individual by reviewing a resume or CV, since they must have at least a phone conversation to judge this quality. The standard one-one-one, inquiry-response interview process does not always allow the opportunity to assess an individual's core values. An optimal interview process should provide the ability to view an individual from multiple perspectives, and in different situations. For instance, when purchasing an automobile, one would not walk onto an auto lot, read their brochure, ask the salesperson questions and then say "I'll take that one!" Not only do you read the literature and ask questions, but you examine the vehicle from different perspectives and under various circumstances. How does it feel when driving it? How does it handle around sharp turns? What are its safety features? When you step on the gas pedal, how well does it respond? The interview process also needs to be varied to learn about the person under *different* circumstances. However, do not turn this into a police interrogation, as the interviewee must also come away with an appreciation for the organization and its people – *and* want to work there!

A standard screening process may start by placing an ad, screening resumes and making preliminary calls to select candidates, narrowing the candidates through questions on the phone and deciding upon who to bring in for an in-person interview. The remaining process is one that we adopted at a start-up biotech company growing from 3 individuals to 25 in a period of about four years.

Have the Interviewee Give a Presentation

When interviewing academic scientists, it is not uncommon to have them give a presentation about some aspect of their work when being interviewed. We also transferred this convention to many of our interviewees, mostly those at a senior level. We ask that they give a 20–30 minute informal PowerPoint presentation about their background and experience, or any particular subject they would like to tell us about. All company employees are invited to listen to the interview presentation and ask questions. We can see, how the candidate handles themselves in front of others, their choice of subject matter, how they attribute credit, and we learn of their visible strengths and weaknesses. A presentation also streamlines the remaining interview time because the candidate only needs to talk about their background once, then later, we can individually focus on specific follow-on questions. All positions do not require professional presentation skills, and a poor presentation does not eliminate the person from consideration, it is just an opportunity to see the individual in a different situation and get to know them better. Usually an interviewee will choose to talk about some of their accomplishments at their current or former employment, and from this we can learn about their ability-to-execute. During one presentation, a candidate chose to include some information about their family and their children's accomplishments in such a way that pride and caring was evident in their personality. This process helps us identify core values, while learning more about a candidate's background.

Face-to-Face Time

After the presentation and Q&A, the interview process continues with many one-on-one or group interviews. Individuals from all disciplines of the company, all ranks above and below the hiring position, interview the candidate. We encourage employes to also share with the candidate what they do and what they think about the company. This can be a true test of an organization's culture if the interviewee still wants to work there after asking questions of employees about what they think of the management and company. Because of the number of interviews, this makes for a longer day, but it provides everyone with an opportunity to ask questions of the potential new team member. This expanded interview process helps with "buy-in" from all employees, and also gives voice to any issues that may come up in the interview process, and the candidate can openly converse with many individuals.

At the completion of each interview, we get written comments from everyone on a single-page preprinted response sheet that can be filled out in 2–3 minutes. This feedback gives us candid comments and a written assessment from everyone. For instance, with one interviewee, several subordinate employees independently commented they found the candidate to be condescending during the interview, whereas senior staff and peer level positions did not observe any such behavior. This alerted us to a potential problem, which we examined further and found to be consistent with other related problems in their previous employment. Don't feel like everyone needs to have the same enthusiastic opinion about the candidate. This process should not be consensus hiring, nor should a company adopt hiring by vote. What is needed, is for all employees to understand that management wants to know any issues of concern before hiring any team member and that they value all input.

Use a Work Behavior Assessment Profile

Later in the interview process, we give the interviewee a work behavior profile to complete. The one we use is the DISC profile; however, a company can use any method that allows them to assess another aspect of the interviewee. We use this tool because it gives us insight into their work behavior and communication style, along with other criteria before we make a hiring decision. Be sure to let the interviewee know there are no correct answers for these profiles, so filling it out in a certain manner will not make any difference. We have had individuals try to guess what we would like to see for responses, but their profiles made no sense. Also, we give the interviewee a copy of their results if they would like one. This assessment is not used for making a hiring decision, it just provides more information on how the candidate views their environment, and it is another factor that helps in communication if they were hired.

Use Informal Meetings

We always include a lunch or dinner with the interviewee and with selected individuals from the company, and in some cases, participation by the interviewee's spouses when applicable. This becomes both a social occasion and another setting to get to know the person better. Discussion usually includes both work and non-work-related subjects. Face it, if you hire this individual you will be spending a lot of time with them, and although they do not have to be your pal, it helps to know them better and to like working with them. When spouses are present, it provides an opportunity to have interaction with more than just the interviewee. Many times spouses volunteer valuable insight about the interviewee, and we get the opportunity to interact with a family member. For senior level hires this is an important facet, because of the large time commitment to a start-up, and it is important that any candidate have support from their spouse. This is a relaxed opportunity get to know them better and to learn more about their core values.

Have Them Talk with a Professional

For executive and senior level positions, when a potential candidate is selected, there is one last discussion with a friendly trained professional, either in person or on the phone. The person we use is, by practice, an industrial psychologist, with a lengthy career in personnel recruiting, who runs their own personnel placement business. This person has personally interviewed over 6,000 individuals and placed the best ones in her client organizations. By having someone with a friendly, conversational style, and a disarming manner, the interviewee can talk candidly, and this allows both parties to discuss the position and its needs. This interview provides more insight into the individual, and helps the individual know more about the position requirements. The questions during this conversation tend to be open-ended, which lets us learn more of their motivations and their potential for contribution to the organization.

Making the Final Decision

Start-up companies do not have lots of time to deal with corrective behavior and remedial problems, so the best defense is a preventative one. Include many different learning opportunities when interviewing individuals for hire in an organization. The interview process should provide the environment in which to identify the most suitable candidates for the position. Whatever methods are chosen, be sure to evaluate all three characteristics of an individual before making a hiring decision.

How to Find These Individuals?

There is a wealth of resources on this subject, and it is not unique to biotech companies, so we will not spend too much time on this subject. If the company is located in a biotech cluster, there should be no trouble finding experienced individuals; however, the challenge may be paying them the salary they command from other well-funded companies. For companies not located in a well-known biotech cluster, the challenge will indeed be finding and recruiting specialized expertise to their locale. It may be very difficult to recruit the type of biotech expertise needed because candidates fear that if the company fails there are no local alternatives for employment. This is not an unfounded concern, so rely heavily on other advantages, such as the high caliber of the science, the great market opportunity for the product, and the large incentives for equity participation in the company. Seek assistance from a local Chamber of Commerce or other civic organizations to show the candidates things that may be of social interest to them or their family. The best candidate to relocate to a nonbiotech locale is one who grew-up, graduated, attended a university, or has family members there. These individuals may have previously left to work at a company in a biotech cluster. It may be possible to identify these individuals through alumni addresses from your local university. Let the university know you are not soliciting alumni for donations, rather seeking potential to bring them back for employment. This may be an excellent way to find stellar candidates. I have personally seen difficult to recruit senior positions filled by individuals who were glad to accept a biotech start-up position at a non-biotech hub, in a place they previously called "home."

When seeking to fill senior executive positions, the importance of good recruiters is understood, only be sure they are well-respected in the industry, and that they do a good job in all aspects of reference checking and support. I have seen recruiters place a VP of Business Development, only to have the company CFO find out a year later, that the employee's MBA was falsified. This issue was only uncovered when the CFO did a education verification because they were suspicious when the VP was unable to accomplish anything of significant value. I know of another situation where a recruiter placed a senior scientific person in a company, only to find out later he had a pending lawsuit from his previous employer for allegations of intellectual property theft. If the company already has venture capital or institutional investors, they may be an excellent resource for finding good applicants or for recommending reputable recruiters. Some VCs may even have a staff member dedicated to recruiting senior level employees for their portfolio companies – this is another added value to a good venture capital partner. Alternatively, your corporate attorney may be well connected, so ask them for help with sources of specialized biotech personnel.

Management Skill Sets

As a company grows, many scientific and technical persons will grow to become managers of others. Unfortunately, not every individual comes with good management skills. An important responsibility of a growing company is to train and equip new managers and leaders. The following is a list of characteristics to instill in all new managers.

Five Skills of Successful Mangers

1. **Ability to Identify:** Ability to recognize the right people for the right positions and projects at the right time.

2. **Ability to Inspire and Lead:** Knowledge of how to inspire and motivate individuals to achieve their goals, objectives, and tasks and to see the vision set before them.

3. **Ability to Instruct:** Ability to teach and instruct individuals to accomplish the objectives that align with the vision the company has created.

4. **Ability to Inspect:** Ability to appropriately monitor and inspect the progress and results from the efforts of these individuals, and to make necessary adjustments in order to continue progress toward those goals.

5. **Ability to Reward:** An understanding of the value of praise, appreciation, increased responsibility, incentives, and remuneration; a different mix is required for different individuals.

If a leader does not possess all of these traits themselves, the first thing they need to do is work on improving their own management skill set, so that they can teach them to others.

Letting Some Go

One of the hardest decisions for a leader is the decision to let an employee go. This is even more difficult when the employee is an early hire or a cofounder of the company. This situation may arise when an employee performs adequately during the early development stage, but later cannot contribute to the growing or changing needs of the company. Sometimes this happens because the individual does not want to make the changes necessary for growth of the organization. Occasionally, these individuals were set-up for failure when they were given Vice President or C-level positions, just because they were cofounders or early hires. Since entrepreneurs have many other things to deal with, personnel issues like these usually are side-stepped because they are challenging to deal with, or because the problem individual has a long history with the organization. Do not be afraid of making tough decisions. Once it becomes clear that this is not a good fit and there has been ample effort to deal with or correct it, the situation must be dealt with quickly and decisively. Sometimes companies procrastinate and choose to deal with the resulting problems symptomatically. This usually means that someone must cover for an individual's shortcomings, or someone always has to double-check their work because of a lack of confidence in their capabilities. This situation reinforces the importance of evaluating carefully whether an individual fits during the formative stages of the company.

If the personnel problem is with an individual in a senior leadership position who consistently cannot perform at the level of their responsibility, weigh the significance of management time, damage to employee morale, and the likelihood of their future value to the organization. Ask yourself, would this person be good for the organization as it grows later? If the answer is clearly "no," then you are setting up the organization for failure by not dealing with this now. There are some alternatives to this situation, but they are related to the capabilities of the problem individual. For instance, if the individual is a great strategy thinker but a terrible manager, then consider a role as a nonmanaging member of the organization. If the individual is creative and provides ideas, but cannot carry out or manage projects, then possibly a consultant role to the organization will work. If the relationship is such that it would preclude any of these or other alternatives, then termination is the only alternative. There are HR issues that will need to have been documented. These include: performance reviews, corrective action, and clear details about their performance and expectations. Other requirements include, documentation of the problems, assistance and help to remediate the problems, and warnings that the lack of correction or improvement may result in termination. These things must be done prior to reaching a termination decision. Most states acknowledge "at-will" employment which just means that all employees can be terminated "at-will", provided there is no discrimination against any member of a protected group of individuals, such as the aged, handicapped, minorities, and various others. It is important to consult with an HR attorney, or if the company has a shared-employer PEO, they are there to provide guidance during this process.

Ignoring a problem employee damages the effectiveness of the organization over time, and diminishes the expectations of other employees, especially if this individual holds a key role in the senior management. Ultimately, the decision should be one that is best for the organization, rather than what is best for one individual. In the StarTrek movie, *The Search for Spock*, the entire Enterprise crew risked their lives to find and rescue Spock. When he was found and returned, Spock logically asked the Captain, "does the good of the one outweigh the good of the many?" In rescue missions and stories of heroics, this is certainly true, but in organizations, particularly start-up and development stage organizations, the good of the one cannot outweigh the good of the many.

Occasionally the "one" may even be the founding CEO. At times, in rapidly growing biotechnology companies, the growth and responsibilities of the organization surpass the abilities of the CEO – or sometimes they are just unwilling to change. There are times when the founding CEO needs to depart for the good of the company. Sometimes this may be a hostile departure, and other times it may be a planned succession. This subject is discussed in more detail in Chapters 14 and 15.

When to Use Separation Agreements

If the problem employee cannot or will not improve, and occupies a senior level position, and the company has decided to terminate the individual, you may want to consider a separation agreement. This document helps the company and the employee part amicably under conditions that satisfy both parties. A separation agreement may contain promises such as, the terminated employee will not sue, file a regulatory complaint, compete, hire employees, or take clients. In exchange, the employer gives something of value. If the employee was a founder or early hire, they may hold large blocks of stock or vested stock options. However, under a previous employment agreement, the company may have already addressed how much they can take with them, but if not, consider giving them an extension to exercise a percentage, but not all of their options. The separation agreement should be structured to deal with all issues related to termination of an individual including, disparaging remarks, disclosure, willingness to sign patent assignments, noncompete and issues with recruiting current employees. In this situation, the company needs to utilize their corporate attorney or an HR attorney to advise them on how to proceed.

Summary

Early hires are critical to getting a solid start on product development and they are significant to raising capital. A company must have notable and credible individuals in all areas where the company professes expertise. To be sure to hire the best-fit individuals, develop an interview process that allows the examination of the three important characteristics of fit: relevant experience, ability-to-execute, and shared core values. Hire the best-fit individuals and have a well thought-out strategy for bringing them on at the appropriate level within the organization. Use a priority listing to time the appropriate hires at a level and capacity that is consistent with the company's changing needs. As the company grows, do not avoid problem employee issues by failing to address them quickly, as they can destroy a team's motivation and limit a company's ability to make timely development progress. Each member of the team will either be an asset or detriment to progress, so identify the best ones and continually help them learn to become better managers and leaders. This requires constantly communicating a vision for the company and incorporating the five skills of effective managers.

Chapter 12
Strategic Alliances and Corporate Partnerships

It is a rare biotech company that can advance their product to commercialization without the help of a corporate partner or strategic alliance. A good partnership can tremendously accelerate product and market development. Not only is a strategic partner advantageous to a development-stage biotech company, but they are almost essential. In this chapter, we discuss strategic alliances, some of the pitfalls, and the value they can bring to the company.

What Are Alliances and Partnerships?

In the biotech industry, we constantly hear of "strategic alliances," "joint ventures," "licensing partnerships," "comarketing agreements," "copromotion agreements," "codevelopment partnerships," and many other types of corporate partnerships. These terms represent various ways relationships can be structured between two or more interested organizations. The purpose of these relationships vary from sponsored research arrangements to much more complex structures such as joint ventures. Alliances and partnerships are typically formed between two or more organizations with differing, but complementary strengths and needs. These relationships are entered into for a specific objective. It is not my intent to define them all, but rather to emphasize the importance and value of finding complementary partnerships for an organization. Partnerships differ from licensing agreements in that partnerships are risk-sharing relationships. Each entity enters into the alliance bearing the risk of the project, and both share in its success or failure.

Why Are Partnerships Important?

For a biotechnology company to be successful, they must capture their market with innovative products before any of their competitors. This requires that the organization be the first to develop an innovative new drug, device, test, or biotech service

C.D. Shimasaki, *The Business of Bioscience: What Goes into Making a Biotechnology Product*, DOI 10.1007/978-1-4419-0064-7_12,
© 2009 American Association of Pharmaceutical Scientists

that provides benefits beyond existing products or services. A small start-up bio-technology company, no matter how capable, does not have the depth of resources and capabilities to reach this goal alone. For the small biotech company, their limited internal abilities can be leveraged from a corporate partner. A good partner can bring product development expertise, clinical trial capabilities, and also improve the success of gaining regulatory approval. A study published in Drug Week concluded that strategic partnerships between a biotech company and a pharmaceutical company resulted in 30% more likelihood of drugs gaining approval from the FDA, compared to those that were developed independently.[1]

What Does a Biotech Have That a Strategic Partner Would Want?

In order to attract a partner, the biotech company must have something of value a potential partner wants. So what does a relatively small development-stage biotech company possess that a large organization would want, which they could not produce themselves? Some of these advantages include:

- The biotech company has specialized research and development capabilities that the larger corporate partner does not have, and this may be of great value for the development of new products in their pipeline.
- The biotech company has technology and patents that the corporate partner does not have which can lead to future products they want.
- The biotech company has a proprietary database and information that the corporate partner wants access to, which they believe will help them develop new products.
- The biotech company has products in their pipeline which the corporate partner wants because these are

  - in a market segment that they already serve
  - in a market segment in which they want to grow

- The biotech company has unrelated products that have tremendous future sales potential.
- The biotech company has related products that the corporate partner can market internationally using their existing infrastructure.

Most biotech partnerships are usually formed between unequal-sized entities. For biotech companies in the small molecule or biologic therapeutic field, including drug delivery, strategic partnerships will usually be formed with larger pharmaceutical companies. For biotech companies in the diagnostic field, partnerships may be more varied, but tend to be marketing alliances with companies that have

---

[1]Drug Development; Pharma-Biotech Alliances Present Lower Risk Opportunities, Drug Week (Biotech Business Week), January 5, 2004.

complementary products in their market. There is increasing interest from biotechnology companies to form alliances with other biotechnology companies; however, there must be complementary benefits in order for these types of relationships to be of any value. Greater innovation is found in smaller entrepreneurial companies compared to larger organizations because larger organizations usually do not support an infrastructure that promotes the type of risk-taking that is essential to innovation. However, smaller organizations do not have the infrastructure and well-honed expertise of product development, regulatory, manufacturing, and marketing that larger organizations have. When any two organization's strengths and needs are complementary, an opportunity for a good alliance arises.

What Does the Biotech Get from a Partnership?

There are tangible benefits a development-stage biotech company gains from a partnership with a large organization having complementary expertise. First, there is cash! Most partnerships formed between large organizations and smaller biotech organizations result in a large cash infusion to the biotech company. To develop any biotech product, it takes lots of cash and time. The right partnership can provide cash and help reduce the product development time. Cash payments to the biotech company can be triggered for reasons such as:

- For the purchase of equity in the company
- As a technology access fee
- At the completion and signing of a collaboration agreement
- As payments upon reaching development milestones
- Upon completing various regulatory filings, clinical trials, NDA, etc.
- Upon regulatory approval for marketing
- Based upon royalties on sales
- Through minimum annual royalties

Although cash to a small biotech organization is its lifeblood, there must be more benefits than cash to make this relationship successful. One important potential benefit to the biotech company in a partnership is the acceleration of their product development. Pharmaceutical companies possess fully integrated clinical trial and regulatory expertise, that a development-stage biotechnology company does not. The list below outlines some of the benefits to the biotech company in an alliance:

- Cash
- Clinical trial expertise
- Regulatory expertise
- Marketing partnerships (domestic and international can be split)
- Manufacturing partnerships

The pharmaceutical company will also derive great benefit. Most mature pharmaceutical companies have limited drug development pipelines, that need to be filled to meet shareholders expectations for growth. Big pharma companies target minimum revenue growth of 5–8% each year. For these mammoth organizations, this incremental growth translates into the need for billions of new dollars in revenue each year. Data from Tufts Center for the Study of Drug Development indicates that in order to reach these revenue goals, pharmaceutical companies would have to get marketing approval for about 2–9 new products each year. This turns the pharmaceutical companies' eyes toward product development opportunities in the biotech industry.

A typical partnership for drug development could include the sharing of the company's development costs, or the larger company taking over the majority of development responsibilities from the smaller company.

Alliances also provide a validation of the biotech company's work. Just as peer-reviewed publications and grants validate the science – good partnerships validate a product and target market potential.

Can't a Potential Partner Just Acquire the Biotech Company?

Yes, a larger organization can acquire the smaller biotech company, and it may even be cheaper in the long-run to do this. However, large organizations such as pharmaceutical companies do not necessarily want the enterprise-risk associated with the company itself, but they are willing to accept an opportunity in the product or technology. One reason for this is that if the product or technology does not work as expected, they can later walk away from the deal. Whereas, if they own the company, and the product is not successful, they will need to continue supporting the organization, or sell it at potentially less value than what they paid for it. These lessons were learned the hard way, such as in 1986 when Eli Lilly purchased Hybritech, a San Diego biotech company for about $500 million in order to get access to monoclonal antibodies and product development for diagnostics and therapeutics. There were culture clashes and an exodus of many employees, several of whom started other biotech companies in the San Diego area. As a result, this acquisition did not meet the expectations of Eli Lilly. Hybritech was later sold to Beckman Instruments in 1995, for an undisclosed amount, but estimated to be a small fraction of the purchase price. For the large partner, forming an alliance and sharing the product development risk is preferred. However, if the product and relationship is successful, many times this relationship will lead to a potential acquisition of the company later.

The Partnership

No matter what stage of product development, a biotech company should start seeking interested partnerships early, even if they are just beginning development of their product. Although a company may not secure a partnership at early stages, by establishing relationships they will find companies that want to follow their progress over time, and these may become potential partners later. In reality, most alliances are formed at a time when significant product development progress has already been made. The exceptions to this usually include biotechnology companies working in unique development areas that desperately need effective drugs, or companies with enabling technologies such as drug delivery, bioinformatics, or development process-related technologies.

Therapeutic and biologic companies may not see real partnership interest in their product until it reaches Phase II or Phase III clinical trials. For diagnostic and medical device companies, partnerships may be plentiful after their product has obtained regulatory approval for marketing. The further along a company's product is in development, the more value the alliance is to a strategic partner. Erbitux, a treatment for metastatic colon cancer was in Phase III clinical trials when Bristol-Myers committed up to $1 billion dollars for an equity stake in ImClone Systems, and up to $1 billion dollars for codevelopment and copromotion rights to their drug. When a future product is in a disease category that is of significant interest to the larger partner, and they lack adequate products for that particular disease, the biotech company may receive interest before they reach clinical testing. Although this is less frequent, it does happen, such as interest in AIDS drugs in the early 1980s, for bioinformatics in the late 1990s, and more recently for Alzheimer's drugs.

What Is the Partnership Process?

First, an organization needs to identify common or convergent interests and goals between their company and any potential partner. In all alliances, the biotech company possesses something of unique interest to the larger partner, and that organization will have something of complementary interest to the company. Early on, the biotech company will be sharing nonconfidential data about their studies that demonstrate some proof of effectiveness. If there is continued interest, the parties will proceed to signing a mutual two-way confidentiality agreement (CDA). The biotech company will then share more detailed development and clinical information with the potential partner, including information about their intellectual property protection. The next step is a face-to-face meeting with the potential partner to share confidential information and discuss common interests to both. If interest remains, there will be a request to send materials, compounds, or prototypes that the larger entity can test in their own laboratories, and put through their own battery of tests for performance. Prior to doing this, the biotech company will need to execute a Material Transfer Agreement (MTA). The MTA protects the biotech company, and limits what the recipient can do, and what they cannot do, with the materials sent. The MTA protects the biotech company by limiting the number of individuals that will have access to the materials, specifying what tests can be performed, prohibiting any reverse engineering of the materials, and requiring that the company provide results of their studies and return any unused materials.

If interest still remains strong, discussions proceed to what type of relationship is sought. This will include a discussion about the expectations of each organization and the proposed terms of the relationship. Sometimes the terms or the type of relationship desired are not acceptable or compatible with the interests of each party, and nothing further develops. Other times, the interest is high, and both parties will negotiate through to acceptable terms of agreement. Once terms are memorialized, then any remaining due diligence will begin. Due diligence assures that the capabilities and resources that were represented exist, and that there are no other issues of concern for the type of relationship desired. Upon completion, a final contract will be signed, and the partnership work will commence. This partnership process is time consuming and does not occur quickly. According to a 2004 IBM Biopartnering survey, the average time taken to complete a bio-partnering deal during 1998 to 2003 was 10.7 months, with about 42% of the deals taking between 11 and 20 months.

Making Alliances Work

In order for a partnership or alliance to work, there must be a clear understanding of the goals and needs of each organization entering into the relationship. Multiple groups will be interacting in a cross-functional manner within the partnership, and there needs to be clear agreement on communication lines, responsible parties, and the objectives of each. The partnership objectives must be clearly defined, and the highest levels of management must be fully engaged to manage and monitor the ongoing progress of the relationship. It is also wise for each organization to select their best and most capable managers to represent the key interactions between the parties. These individuals should be problem solvers and have the flexibility to seek creative solutions to any issues that arise.

As companies enter into an alliance, there is excitement about the partnership, the work, and the opportunity. Corporate attorneys will have drafted an agreement attempting to spell out every type of situation that may possibly be encountered in addition to outlining the responsibilities of each party. Most of the contract language will address each organization's responsibilities, termination reasons, arbitration, and dispute resolution mechanisms as well as a litigation jurisdiction. In other words, the contract is usually written assuming that the partners will *not* do what they should do to make the partnership work – the "worst case" scenario. Unfortunately, this is how contracts must be written – presuming one party will not do its part to meet its obligations. If, during the relationship, a company is having to continually take the contract out of the file to find out what the other is supposed to be doing – they already have a problem. Ideally, contracts should never need to be pulled out again until the project is completed and successful. If the companies find themselves (or their corporate attorney) constantly reviewing the contract to see if the alliance partner is performing their responsibilities, it means that there are problems which must be dealt with, and issues that need to be resolved quickly, or risk failure of the partnership.

Why Do Alliances Fail?

Various studies of all corporate alliances show that slightly over 50% of alliances are successful. Some individual companies may have high alliance success rates of over 80%, whereas other companies may have success rates below 20%. There are many reasons why alliances fail. Some fail because there was never a good fit to begin with, or there was not true agreement amongst the teams that the partnership was beneficial.

Many times companies enter into a relationship without considering all the issues that impact alliances and partnerships. A 2006 biopartnering survey was conducted by IBM comprising 235 companies with 74% being biotechnology, and 26% being pharmaceutical companies, with 52% being US-based, and 48% in other countries[2]. This survey showed that the two main factors companies took into account when entering into alliances were: the deal or the offer, and the scientific expertise. Yet the survey found that culture and communications, proved to be far more accurate in predicting the success of the alliance. They also found that the quality and consistency of the partner's staff, the degree of cultural compatibility between the two companies, and the sponsor's willingness to discuss problems, played the biggest role in making the alliance work (Fig. 1).

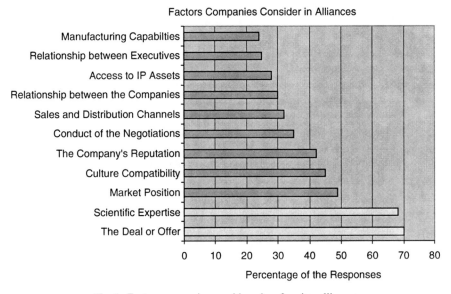

Fig. 1 Factors companies consider when forming alliances

Because of the vast amount of time and effort that goes into forming an alliance or partnership, you should answer several questions such as these before proceeding:

- What are the expectations of each organization?
- Do the companies have shared goals?
- Is there a strong commitment to this relationship at the highest level of each organization?
- Are the strengths complementary?
- Are there good working relationships between the members of each team?
- Are the company corporate cultures similar?
- Is there open communication and a willingness to address problems directly?
- Are both parties seen as valuable contributors to the relationship or is one just a source of a product?

Before entering into any relationship, be sure that the potential partner has the necessary skills, capabilities, communications, and operating culture that match your organization. It is never good to enter into an alliance simply because a company has something of value, but no one particularly likes to work with them; this is a signal that the relationship will be a disaster, or at best, frustrating for both parties.

Summary

Good partnerships and alliances are essential for a biotech company to be successful in developing their products through to commercialization. In the end, alliances are just people working with other people. There is a high likelihood that a partnership will be successful if the motivations are synergistic, there are good communications, the working relationship is valued, and each party addresses issues as they arise. Strong and successful partnerships fuel the formation of other partnerships, and it is not uncommon for a successful biotech company to have multiple alliances rather than just one. Aside from the obvious synergistic benefit of accelerating product development or leveraging a marketing relationship, good partnerships can lead to a future acquisition of the biotech company which provides an exit for its shareholders. By finding complementary and synergistic partnerships, a small biotech company can leverage their partner's strengths and thus help ensure their own future success.

Chapter 13
The Regulatory Process for Biotech Products

A sobering reality is that a biotech company can raise all the capital they need, possess the strongest patent portfolio possible, successfully complete a lengthy product development process, and identify a lucrative market for their product, but if they do not meet regulatory requirements and gain regulatory approval – they will never see the fruit of their labor. Whether it is the regulatory authority in the US (FDA), the European Union (EMEA), Japan (PMDA), or any other country – regulatory approval of a biotech product is one of the most important milestones for any biotechnology company. For companies considering marketing in the US, the Food and Drug Administration (FDA) does not grade on a curve, so it is critical to have thorough knowledge of the regulatory requirements and hurdles for your product's approval. Oftentimes, regulatory planning is presumed to be straight forward and so far into the future that a young biotech company may ignore this until later – this thinking can be disastrous. A thorough understanding of the regulatory process is vital early when charting a development path because it impacts developmental and testing decisions along the way. Whether it is realized or not, regulatory requirements are dynamic and evolve, so a company must constantly have current regulatory knowledge about the issues they will face for new product entrant approval. Also, having this knowledge will improve their ability to raise capital; knowing regulatory success measures allows one to write a better business plan, and gives assuring responses to potential investors.

In this chapter, we review the FDA regulatory process for biotech products and discuss some key points to help navigate through this maze. Most biotechnology products are marketed internationally, therefore it should be realized that each country has its own product approval process, and companies must meet these individual requirements also. This chapter is not meant to be a complete treatise of regulatory issues in biotechnology, but it provides an overview of the regulatory process, and discusses a few issues relevant to product approval. More detail about the regulatory process can be obtained directly from FDA guidance documents or from the FDA website.[1]

---

[1] http://www.fda.gov

C.D. Shimasaki, *The Business of Bioscience: What Goes into Making a Biotechnology Product*, DOI 10.1007/978-1-4419-0064-7_13,
© 2009 American Association of Pharmaceutical Scientists

Although the entrepreneur will not be the sole person ultimately responsible for regulatory compliances, in the beginning they still need to be knowledgeable of these regulations, in order to make wise product development decisions and to properly assess the qualifications of any regulatory expert. Talk with those in the industry to be sure and understand each step in this process and any potential regulatory issues for your product. Most importantly, learn the success measures the FDA and other regulatory agencies use for gauging progress at each of these stages, and ultimately, to final product approval.

Missteps in the regulatory process can cost a company millions of dollars, derail alliances, and cause years of delays. Careful regulatory planning and sound strategy development during this stage is of critical importance. Regulatory missteps are too common in biotechnology companies. Aviron, a start-up biotechnology company in Burlingame, CA, was the original makers of FluMist, the first intranasal vaccine delivery system with the company's first product being a vaccine for influenza. As with all developing organizations, issues arise during manufacturing that need to be resolved or processes need to be improved in order to produce a product in volumes large enough for clinical studies. The company made changes or improvements to the manufacturing process; however, the clinical trial material was produced at a facility different from the one where it was being manufactured for commercialization. During the regulatory filing, the FDA informed the company of the problem, and Aviron had to repeat human clinical trials with material used from their new manufacturing facility. This situation cost the company many millions of dollars by repeating clinical studies, and years were added before they ultimately reached FDA approval and commercialization.

Alphabet Soup

Brace yourself – the government has acronyms for everything. For those not familiar with the regulatory acronyms pertaining to their product, it may seem as if a secret decoder ring or a handy reference book is required to understand what is being communicated. It may sound as if regulatory persons speak in a foreign language because the number of acronyms routinely used within any regulatory agency is mind-boggling. Consider this acronym-filled comment: "The FDA has recently changed its guidance for IVDMIA's pertaining to enforcement discretion and requirements for filing PMAs and 510(k)s, please contact the CDRH's, OIVD, or DSMICA for more information." It is essential to become familiar with the regulatory acronyms used in your field.

Most all the regulatory citations pertaining to biotechnology product approval are contained within Title 21 of the Code of Federal Regulations, abbreviated as "21 CFR." These are rules, sometimes referred to as "administrative law," and are published in the Federal Register by US Executive departments and agencies of the Federal Government. There are 50 different titles. References to individual regulations are stated in the form "*21 CFR 312.20*," which refers to Title 21 of the CFR, part 312, section 20. All these regulations are accessible on the FDA website.

Biotechnology Regulators: The Food and Drug Administration

For most companies developing a biotechnology product for US approval, the FDA will be the federal agency that will review and regulate their product, and will be the Agency determining its fate at the final stage prior to reaching the market. In some instances, such as for clinical laboratory tests, the Clinical Laboratory Improvement Amendment (CLIA) which operates under the Center for Medicare and Medicaid Services (CMS), will be the guiding regulation. Depending upon the product, the United States Department of Agriculture (USDA) may be a regulator, or the Nuclear Regulatory Commission (NRC) may also have a part in the regulation of a product.

The FDA of the United States is still considered by the world to be the premier regulatory agency. The FDA is responsible for ensuring that products that reach the US consumer are safe and effective, and consistent with their labeled claims. The US certainly has had its problems with the contamination of food such as *E. coli* and Salmonella, in addition to many drug recalls. There have also been fatalities from adverse effects of drugs that have reached the US market which were not reasonably safe for everyone. However, the Agency has done a tremendous job when one considers the enormity and breadth of issues they face, and the advances in technology that they must review. We cannot forget that the companies themselves must first bear a heavy burden for these missteps as they may have ignored or failed to implement and monitor safe practices in the process of developing and testing their products – in some cases they may even covered up these issues. If all companies were as safety conscious as if they administered their drug, sold their food, tested their product, on themselves and family members; theoretically, the FDA's job may not be necessary, and more products would be safer with much fewer recalls.

Due to the mounting challenges in product and food safety attributable to both the industry and the FDA, the Agency out of necessity, appears to have adopted a heightened "risk averse" culture related to regulation and approval. Risk aversion occurs when the misdeeds of a few create extreme consequences for everyone. A good example of this has occurred in airport security. Because of the nefarious deeds of a few, the entire public must become suspect and scrutinized to a level not normally considered rational – such as taking away a fingernail clipper from a 92-year-old grandmother because she could use it as a weapon to highjack a airplane. The point is, that everyone becomes suspect and the worst is (and must be) presumed of all because of the misdeeds of a few. Thresholds for safety are continuously raised because of the potential downside risks. As safety thresholds become higher and technology rapidly advances in the biotech industry, it is challenging to quickly move advanced technology products through the FDA in a timely manner. Be aware of this challenge when estimating the timeframe for regulatory approval of any new product. Always anticipate that what seems to be straightforward to the company, may not be viewed the same way by regulators.

What is an Institutional Review Board?

Before a company can conduct any research involving humans or human specimens, they must first get their study and their investigators approved by an Institutional Review Board (IRB). An IRB is a group of individuals formally designated to review, monitor, and approve medical research protocols. The IRB policy was established for the purpose of protecting the rights and welfare of human subjects. In the US, the FDA and the Health and Human Services (HHS) department empower IRBs to approve, require modification of, or disapprove proposed research.

Larger medical or research institutions such as universities, academic institutions, and hospital systems have their own IRBs. Smaller organizations and companies usually do not, though they can constitute one if they so desire. Independent IRBs are available to fill this void. These contract IRBs are used heavily by small or emerging biotechnology companies that do research and clinical studies involving humans or collect human specimens. Since independent IRBs are also commercial enterprises, they usually have quicker turnaround times than do institutional IRBs. Contract IRBs may hold meetings once or even twice weekly, whereas most institutions hold monthly or even quarterly IRB meetings. The value to a contract IRB is that a protocol can be approved or amended more quickly, and the company can get started sooner on their study. IRBs are comprised of scientific, medical, and nonscientific and nonmedical individuals, who examine the ethical nature of the research. Many times these nonmedical or nontechnical positions are filled by clergy, attorneys, or lay persons. The underlying function of the IRB is to protect the rights of the patient, and this is done by insuring that the research is ethical, and conducted with the patient's informed consent. The IRB focuses heavily on the on the language that is contained within the informed consent.

Informed Consent

The safety and ethical nature of clinical testing is carried out by obtaining Informed Consent from subjects before any testing or research is performed on the patient, or before collecting any specimen for study. Informed Consent is ensured by the signing of this legal document outlining to the patient the reasons for participating in the study, and the risks associated with participating. There are limitations on the age at which Informed Consent can be given. Most clinical studies involve some form of compensation to the patient for participating, which can be monetary or in-kind medical services. The limitations of compensation are determined by the IRB, as is the manner of advertising and recruiting for patient enrollment. The IRB is careful to ensure that the sponsoring organization does not entice the patient to the point where they cannot make a rational decision about participating in the study

Regulating Drugs and Biologics: The Drug Approval Process

The FDA regulates drugs and biologics through two divisions, the *Center for Drug Evaluation and Research* (CDER) and the *Center for Biologics Evaluation and Research* (CBER). The approval process is similar for both, but there are variations in requirements. Depending upon the type of therapeutic a company is developing, they will be interacting with one or the other of these divisions.

Pre-IND Meeting

As discussed in Chapter 6, it is a good idea to hire a regulatory consultant experienced in the treatment indication you seek prior to any discussions with the FDA. Before filing an Investigational New Drug Application (IND), the company will have had a Pre-IND meeting with the FDA to discuss their clinical research plans, their proposed endpoints, and what they anticipate accomplishing during their clinical studies. They will then hear from the FDA about their views, and what they would like to see from such studies. This is an opportunity to clarify what endpoints the FDA will be looking for, and what the company needs to accomplish in order to satisfy concerns about product safety and efficacy. Be thoroughly prepared before going into an IND meeting, as you can get entirely different answers to the question "what studies would the FDA like to see?" versus the question "what does the FDA think of this plan?" During the IND meeting, bring key people that will be supporting the clinical testing. These may include a regulatory person, a medical director, a clinical trial manger, a biostatistician, the most senior scientific person, and a businessperson responsible for the company direction.

When outlining a Phase I study plan, be sure to simultaneously view the entire clinical testing goals and objectives, rather than considering Phase I, II, and III as independent studies. The end result is product approval; so build into each of these study phases, measures that generate critical information to ultimately support approval, rather than just support the next phase of testing. As a company considers study endpoints, they need to think about their indication-for-use strategy and their ability to achieve the required endpoints. For diseases that have early, mild, moderate, and severe stages of a disease, this may pose a greater challenge in demonstrating product efficacy. It may not always be the best decision to tackle the most difficult indication-for-use first. Sometimes, by first targeting a less-risky indication that still has significant market value, a company may be rewarded with a higher certainty of achieving an endpoint that reaches statistical significance.

Investigational New Drug Application

This is the first formal step in the process toward commencing human clinical studies (see 21 CFR part 312). The IND permits the sponsor (company) to submit to the FDA, data supporting the reasons why their drug candidate should be moved from animal testing into the first phase of testing in humans. In legal terms, the IND is really an application requesting exemption by the sponsor from the Federal law that requires a drug approval before shipping it across interstate lines. US Federal law prohibits shipping any unapproved drug across interstate lines, and an IND is a waiver request to the FDA for exemption from this legal requirement. In practicality, the IND is the means by which the FDA determines if the drug is safe to begin clinical testing in humans, and if your investigators are qualified to perform these studies.

Companies testing a small molecule drug or similar compound will file their IND with the CDER. If the product is a biologic or vaccine, the company will file their IND with the CBER. Within the IND, the sponsor has substantiated that the new drug candidate or biologic has reasonable safety and efficacy data to support the movement into human testing. Significant volumes of supporting information are submitted to the FDA which includes data from these three categories:

1. **Animal pharmacology and toxicity studies.** This data supports the notion that the drug candidate or biologic is reasonably safe for initial testing in humans. It describes the potential toxicity and adverse effects expected during the usage of the therapeutic or biologic. Data from *in-vitro* and *in-vivo* studies are submitted, which usually include two different species of animal models. This section contains all data known as Adsorption, Distribution, Metabolism, and Excretion information (ADME).

2. **Manufacturing information.** This is information that pertains to the composition, manufacture, stability, and controls used for manufacturing the drug candidate. This section supports the company's ability to adequately produce and supply consistent batches of the drug candidate. This information also describe the synthesis process, or the biological production process, and the ability to determine purity and detect contaminants in the process.

3. **Clinical protocols and investigator information.** Protocols and proposed clinical studies are provided to assess whether the trials will expose the participants to unnecessary risks. Included is background information on the investigators and those who will oversee the administration of the drug, in order for the FDA to determine their qualifications for conducting this study. The IND submission will include the dosage(s) used during the study and the number of subjects or patients expected to be enrolled in the study.

After filing an IND, the FDA has 30 days to review or contact the sponsor about its suitability. If the 30-day review period has expired, the sponsor can begin the clinical trials if the FDA has not contacted the Sponsor and placed the application on Clinical Hold. A Clinical Hold is the mechanism by which the FDA can delay a decision if they do not believe or cannot confirm that the study will be conducted without unreasonable risks to the patient. Once a Clinical Hold is issued, the Sponsor must address all issues before the Clinical Hold can be removed. This process may be brief or it may cover an extended period of time, depending on the quality of the data submitted and the risk involved with the testing.

Phase I Testing

Phase I is the initial introduction of an investigational new drug or biologic into humans. After the 30-day IND review period, or after the Sponsor has satisfied any FDA issues that placed the review on Clinical Hold, the Sponsor may begin the Phase I testing of their investigational drug in humans. Phase I testing is typically a small study that is closely monitored, and usually conducted on healthy volunteers, but occasionally conducted on actual diseased patients depending on the type of therapeutic and indication the Sponsor is seeking. The main objective is to determine the safety profile of the treatment, along with determining the metabolic and pharmacologic actions of this treatment in humans. These studies are designed to identify any side effects associated with increasing doses, and, if possible, gain early evidence on effectiveness. During Phase I, sufficient information about the drug's pharmacokinetics (pK) and pharmacological effects or pharmacodynamics (pD) are collected which will permit the design of well-controlled, scientifically valid, Phase II studies. Phase I studies also evaluate drug metabolism, structure–activity relationships, and the mechanism-of-action in humans. During Phase I studies, The FDA can impose a Clinical Hold (i.e., prohibit the study from proceeding, or stop a trial that has already started) for safety issues that may arise during testing, or because of a sponsor's failure to accurately disclose the risks of study to investigators. The total number of subjects included in Phase I studies varies with the drug, but is generally in the range of 20–80, but can be smaller for certain rare diseases or conditions.

Phase I clinical trials will help a company determine the merit of proceeding forward into subsequent stages of human testing where significant expenses will be committed for many more years.

Phase II Testing

Phase II human testing are early controlled studies conducted to obtain preliminary data on the effectiveness of the drug or biologic for one or more indications in patients with the disease or condition studied. This phase of testing also helps determine the common short-term side effects and risks associated with the drug. Phase II studies are typically well-controlled, closely monitored, and conducted in a larger group of patients, usually involving 100–300 patients with the disease or condition the drug will be treating. However, some phase II studies may be much smaller depending on the proposed drug indications and the patient availability. For instance, a treatment for end-stage brain cancer will not be recruiting the same number of patients that a treatment for allergic rhinitis would, nor will the speed of recruitment be the same. Phase II studies are usually placebo-controlled, randomized, and double-blinded (where the investigators themselves do not know who is receiving the treatment or placebo, and the patients are randomly assigned by computer to receive the treatment or placebo). Phase II studies will help determine the optimal dosing for administration of the treatment or biologic. Phase II studies are also referred to as "Proof-of-Concept" studies, which when successful brings greater assurance that the treatment may be effective in treating the disease or condition.

Phase III Testing

Phase III studies are also called "Pivotal Studies" because the success of the drug or biologic is determined by the outcome of these studies. These studies are performed once preliminary evidence has been obtained that suggests effectiveness of the drug in Phase II. Phase III studies are intended to gather definitive information about effectiveness and safety which is needed to evaluate the overall risk-benefit relationship of the drug or biologic. Phase III studies also provide the basis for extrapolating these results to the general population, and this information will then be included in the physician labeling. These studies usually include several hundred to several thousand people, depending on the disease or condition the Sponsor seeks to treat. The number of patients enrolled should be estimated using many sources, including an epidemiologist and biostatistician, to ensure there is enough predictive power to show a certainty of statistical significance. Sometimes clinical study outcomes do not reach "statistical significance" yet they show an effect of the treatment. This could mean that the study did not have enough patients to determine whether or not the effect was a meaningful result. Because of this possible outcome, do not try to save money by shorting the recruitment of enough patients during enrollment. The FDA will give you input on minimum study size and this will be based upon the number of indications you will be seeking for approval.

Throughout the Phase III study, it is critical to carefully monitor that investigators strictly follow the inclusion criteria for patients enrolling at each study center. Once Phase III studies are completed and a product approval application is submitted to the FDA, the FDA can exclude enrolled patients if they do not meet the precise inclusion criteria. If investigators did not screen patients well enough before entry, a study may have reached sufficient patient enrollment numbers, but if the excluded patients are in large numbers, the study may no longer be able to demonstrate statistical significance with the reduced population. Also, as in Phase II, the FDA can impose a Clinical Hold prior to starting if a study is unsafe (as in Phase I), or if the protocol is clearly deficient in design to meet its stated objectives.

As discussed in Chapter 6, patient recruitment can be challenging because of the difficulty in achieving the recruitment goals in a timely manner. Be sure to plan for recruitment delays and do not overestimate recruitment capabilities. Be sure to monitor adherence to patient recruitment criteria from all recruitment centers. It would be disastrous to find out during regulatory review that a good portion of these patients did not meet enrollment criteria and the data package fell short of the numbers to reach statistical significance.

Phase IV Testing

Phase IV studies are also referred to as "Postmarket Studies" where the product is evaluated in studies after it has been approved by the FDA. Sometimes companies are required to conduct these studies to find out about long-term risks and benefits of the product, or to test it in other populations such as children. These requirements are becoming more frequent for products receiving FDA approval.

Pre-NDA Meeting

At the conclusion of the Phase III studies, the sponsor will meet with the FDA to discuss the presentation of data (both paper and electronic) in support of the application. This meeting is called a Pre-New Drug Application meeting (Pre-NDA) or pre-BLA meeting. The information provided from the sponsor at this meeting usually includes:

- A summary of clinical studies to be submitted in the New Drug Application (NDA) or Biologics License Application (BLA)
- The proposed format for organizing the submission, including methods for presenting the data
- Any other information that may be required for your product application review

The Pre-NDA meeting is conducted to uncover any major unresolved problems or issues prior to submitting an NDA or BLA. The purpose of this meeting is to: review studies the sponsor is relying on to determine if they are adequate and well-controlled in establishing the effectiveness of the drug, to help the reviewers to become acquainted with the general information to be submitted, and to discuss the presentation of the data in the NDA to facilitate its review.

New Drug Application or Biologics License Application

The company will file an NDA or BLA with the FDA at the conclusion of their clinical trials, provided these studies show that the drug or biologic is safe and effective for its intended use. The NDA is a comprehensive application with data summarizing all the information from many years of clinical trials and safety testing.

In the NDA and BLA, the company also must demonstrate the ability to manufacture three consecutive lots of the compound or biologic, showing that these products are identical. The final NDA or BLA application is enormous, and it would not be unusual for an NDA to be over 100,000 pages long.

Once the application is filled, the FDA then reviews all the data provided, and will request additional information in support of the application; this process can take approximately 18 months for review. A meeting between the FDA and the sponsor may also take place 90 days after the initial submission of the application in order to discuss issues that are uncovered in the initial review. If the FDA believes that additional experts should participate, a review of the NDA will include an evaluation by an advisory committee, which is an independent panel of FDA-appointed experts, who will hear and review data presented by the company and the FDA reviewers. The FDA uses advisory committees to obtain outside advice and opinions from expert advisors so that final agency decisions will have the benefit of wider national expert input. The FDA will seek a committee's opinion about a new drug, a major indication for an already approved drug, or a special regulatory requirement being considered such as a boxed warning in a drug's labeling. Advisory committees may also advise the FDA on necessary labeling information, or help with other things such as guidelines for developing particular kinds of drugs. They may also consider questions such as whether a proposed study for an experimental drug should be conducted, or whether the safety and effectiveness information submitted for a new drug is adequate for marketing approval. At a special meeting, the advisory committee members will vote on whether to recommend that the FDA approve the company's NDA application, and under what conditions. Committee recommendations are not binding, and the FDA is not required to follow the guidance or recommendations of the Advisory Committee, but more often than not they typically follow their recommendations.

At the conclusion of the FDA application review process, the FDA can either (1) approve, (2) send the company an "approvable" letter requesting more information or studies before approval can be given, or (3) deny the approval. Based upon the conclusions of this review, the company may be required to continue Postmarket Surveillance Studies (Phase IV studies) to monitor and study additional issues that may be warranted. Once the product is approved, the company still has strict reporting requirements about adverse reactions and deaths that must be reported to the FDA during marketing of their new drug or biologic. A schematic of the drug approval process is depicted in Figure 1.

Other Types of Regulatory Review Processes

Accelerated Development Review

There are additional pathways for regulatory approval available for certain products. Accelerated development/review is a highly specialized mechanism for speeding the development of drugs that promise significant benefit over existing therapy for serious or life-threatening illnesses, for which no therapy exists. This process incorporates several novel elements aimed at making sure that rapid development and review is balanced by safeguards to protect both the patients, and the integrity of the regulatory process. Accelerated development/ review can be used under two special circumstances: when approval is based on evidence of the product's effect on a "surrogate endpoint," and when the FDA determines that safe use of a product depends on restricting its distribution or use. A surrogate endpoint is a laboratory finding or physical sign that may not be a direct measurement of how a patient feels, functions, or survives; but it is still considered likely to predict therapeutic benefit for the patient. The fundamental element of this process is that the sponsor must continue testing after approval to demonstrate that the drug indeed provides a therapeutic benefit to the patient. If not, the FDA can withdraw the product from the market more easily than usual.

Parallel Track

Another mechanism to permit wider availability of experimental agents is the "parallel track" policy developed by the US Public Health Service in response to AIDS. Under this policy, patients with AIDS, whose condition prevents them from participating in controlled clinical trials, can receive investigational drugs shown in preliminary studies to be promising. See the FDA website for more detailed information.

Treatment IND

Treatment INDs are used to make promising new drugs available to desperately ill patients as early in the drug development process as possible. The FDA will permit an investigational drug to be used under a treatment IND if there is preliminary evidence of drug efficacy, and the drug is intended to treat a serious or life-threatening disease, and there is no comparable alternative drug or therapy available to treat that stage of the disease in the intended patient population. In addition, these patients are not eligible to be in the definitive clinical trials, which must be well underway, if not almost finished. An immediately life-threatening disease means, a stage of a disease in which there is a reasonable likelihood that death will occur within a matter of months, or in which premature death is likely without early treatment. For example, advanced cases of AIDS, herpes simplex encephalitis, and subarachnoid hemorrhage are all considered to be immediate life-threatening diseases. Treatment INDs are made available to patients before general marketing begins, typically during Phase III studies. Treatment INDs also allow FDA to obtain additional data on the drug's safety and effectiveness.

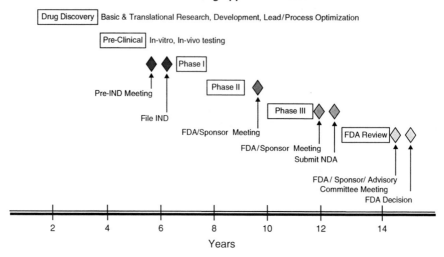

Fig. 1 Drug development and approval process

Regulating Medical Devices, IVDs, Diagnostics and Laboratory Tests

The regulation of Medical Devices falls under the authority of the Center for Device and Radiological Health (CDRH). If a company is not sure about where their product may be categorized, the first thing to do is check whether the product is defined as a medical device by the FDA. The FDA defines a Medical Device as "an instrument, apparatus, implement, machine, contrivance, implant, in vitro reagent, or other similar or related article, including a component part, or accessory which is:

- Recognized in the official National Formulary, or the United States Pharmacopoeia, or any supplement to them

- Intended for use in the diagnosis of disease or other conditions, or in the cure, mitigation, treatment, or prevention of disease, in man or other animals

- Intended to affect the structure or any function of the body of man or other animals, and which does not achieve any of its primary intended purposes through chemical action within or on the body of man or other animals, and which is not dependent upon being metabolized for the achievement of any of its primary intended purposes

Many products are considered to be Medical Devices by the FDA which may not be generally thought of as medical devices. Items such as tongue depressors and bedpans are Medical Devices and regulated by the FDA. More complex apparatus such as pacemakers with micro-chips, laser surgical devices, and X-ray machines are also Medical Devices, and are more readily understood to be medical devices. *In-vitro* diagnostic products such as: general purpose lab equipment, reagents, and test kits, which include monoclonal antibody technology, are considered to be Medical Devices by the FDA. In addition, some laboratory testing services that use certain reagents are considered Medical Devices.

Laboratory testing services that produce reports based upon multianalytes and interpreted by complex algorithms are now considered to be Medical Devices, and the FDA has recently established a new class of products termed, In Vitro Diagnostic Multianalyte Index Assay (IVDMIA). These are now a subcategory of tests that were previously known as "home-brew" or "analyte-specific reagents (ASR)." There is still great uncertainty regarding how these tests will be phased into FDA regulations since they are services, rather than tests kits or consumables, and these tests are currently regulated by CLIA. It will be quite challenging to harmonize additional regulatory requirements from the FDA because IVDMIAs already have existing CMS regulatory and state licensure requirements.

Medical Device Classification

Once it is determined that a product is a "Medical Device" as defined by the FDA, the next step is to determine the classification in which the medical device belongs. There are three classifications possible for a medical device, and all products will fall into one of these. You guessed it – the Classifications are Class I, Class II, and Class III. Each of these three classifications has certain requirements for regulatory compliance, and they have to do with the safety and type of controls placed on each classification.

1. *Class I devices*. Devices for which general controls of the Act are *sufficient* to provide reasonable assurances of their safety and effectiveness. They present minimal potential for harm to the user and the person being tested. Most Class I devices are exempt from premarket notification.

2. *Class II devices*. Devices for which general controls alone are *insufficient* to provide reasonable assurances of their safety and effectiveness, and for which establishment of special controls can provide such assurances. Special Controls may include special labeling, mandatory performance standards, risk mitigation measures identified in guidance, and postmarket surveillance.

3. *Class III devices*. Devices for which insufficient information exists to provide reasonable assurance of safety and effectiveness through general or special controls. Class III devices are usually those that support or sustain human life, are of substantial importance in preventing impairment of human health, or are devices which present a potential, unreasonable risk of illness or injury.

The Regulation of medical devices is handled by the Office of Device Evaluation (ODE) which has five divisions:

- Division of Cardiovascular Devices
- Division of Reproductive, Abdominal, and Radiological Devices

- Division of General, Restorative, and Neurological Devices
- Division of Ophthalmic, and Ear, Nose, and Throat Devices
- Division of Anesthesiology, General, Hospital, Infection Control, and Dental Devices

In Vitro Diagnostic Devices

The In Vitro Diagnostics (IVDs) are separately worth noting, and these are a subcategory of Medical Devices as defined by the FDA because it meets the definition under the Federal Food, Drug, and Cosmetic Act, Section 201(h). Most other Medical Devices function on, or in a patient, whereas IVDs include products used to collect, prepare, and examine specimens (e.g., blood, serum, urine, spinal fluid, tissue samples) *after* they are removed from the human body. The FDA divisions that are responsible to review IVD products fall under the following two centers, and these are the groups a company will be communicating with during their product's regulatory review.

1. **Center for Devices and Radiological Health (CDRH)**

    (a) Office of In Vitro Diagnostic Device Evaluation and Safety (OIVD)

    - Division of Chemistry and Toxicology Devices
    - Division of Immunology and Hematology Devices
    - Division of Microbiology Devices

2. **Center for Biologics Evaluation and Research (CBER)**

    (a) Office of Cell, Tissues, and Gene Therapy (OCTGT)
    (b) Office of Blood Research and Review (OBRR)

    - Division of Blood Applications (DBA)
    - Division of Emerging and Transfusion Transmitted Diseases (DETTD)
    - Division of Hematology (DH)

Investigational Device Exemption

Before beginning clinical testing with a medical device or product, the company needs to determine if preapproval is required to begin clinical studies. In some cases, more than one branch may review a product, particularly those that are drug/device combinations. Preapproval is required of Medical Devices that are categorized as Significant Risk Devices. If a product is a Significant Risk Device, the company will need to file an Investigational Device Exemption (IDE) application with the FDA prior to testing their device in humans, or on clinical specimens obtained from humans. Your IRB will help to determine whether a test or device is classified as a Significant Risk Device. For nonsignificant risk (NSR) device studies, an IDE is considered "approved" when a sponsor meets the abbreviated requirements found in 21 CFR 812.2(b), which includes approval from the reviewing IRB. The IDE is an application which, when approved, allows the device to be shipped lawfully for the purpose of conducting studies regarding the safety and effectiveness of the device, without complying with certain requirements of the Federal Food, Drug and Cosmetic Act (FD&C Act).

Division of Small Manufacturers, International and Consumer Assistance

Within the CDRH there is a Division of Small Manufactures, International and Consumer Assistance (DSMICA) that has been specifically set up to help relieve the frustration, and avoid the confusion that comes from working with an enormous agency such as the FDA. This resource is helpful for small manufacturers, so be sure to make use of these resources which the federal government has provided to assist with the regulatory process.

Medical Device Approval Process and Filing Route

Once clinical studies or clinical testing is complete, the company will file an application for marketing clearance or approval. Each application is different and has different requirements for approval. The company will have previously determined with the FDA, which marketing approval process the product will be required to meet in order to go to market. There are three major routs for a Medical Device:

1. *Exempt device* which means that a product is exempt from filing an application for marketing clearance or approval, but still has certain requirements to meet prior to commercialization, which are minimal.

2. *510(k) or Premarket Notification* is the process named for the section of the Current Federal Register (CFR) cited. A 510(k) is an application for marketing clearance and not technically an "approval" but rather a "marketing clearance." Regardless, a company still cannot market their product without the FDA's clearance of a 510(k). The 510(k) has a "substantial equivalence" requirement where the product is allowed to be marketed *if* it is demonstrated to be substantially equivalent to a pre-1976 amendment device. The FDA has 90 days to review the application, but realize their clock will stop when they request more information or have questions. If there are significant issues, the clock can in some cases be reset. Generally, if you have followed the FDA guidance and do not have a completely new category of product, the timeframes for marketing clearance are relatively short.

3. PreMarket Approval (PMA) is the most arduous and lengthy approval process for a Medical Device. It is almost the equivalent of a drug approval process for a Medical Device, and in some cases the review process can take almost as long. There are filing mechanisms that the FDA has in place that allow a company to complete this in a modular format to accelerate the process for a PMA. Medical

devices that require a PMA are those that are deemed to potentially be a significant risk to the health and safety of the public. Though the FDA has 180 days to review this application, in real time, the application process can average 18 months for approval. In addition, there is a higher hurdle for approval as compared to the 510(k) process.

Know that the regulatory review timetable is variable, and it depends on the complexity of the product and quality of the supporting data. While I was working on infectious disease diagnostics at one start-up company, we took five products through the FDA 510(k) process that were based upon new technology. The total review time was extremely variable, with the shortest review taking 90 days, and the longest review taking 17 months before receiving marketing clearance.

Remember that Regulators are First People

There are valid reasons to have different points of view with the FDA on issues related to regulatory approval requirements and interpretation of data. When there appears to be irresolvable differences, the FDA has an ombudsmen to assist in such differences when they cannot be worked out amicably between the FDA and the sponsor. However, a company will have better cooperation, and the FDA may be more willing to listen, when you approach reviewers as you would any other colleague you work with. Regulators are still people, and as such, they can make mistakes, misinterpret information and communications, and forget. By working with them in a manner that takes into account the high stress level and the significant responsibility and accountability of their position, one may find that the reviewers will respond better, and the working relationship will be more collegial. This is not to say that by having a kinder, gentler approach with the FDA your company will get their product approved, and the FDA will accept everything that is submitted. However, it may give the company the benefit of a doubt when there are issues that involve discretion, and it may help expedite some parts of the process. Let the FDA reviewers know when they do a good job, and they may also be willing to listen when there are things that they can improve upon. The regulatory review process is challenging enough, but by having a good working relationship with the FDA staff, a company can make this process less stressful, and possibly enjoyable at times. Do not forget that the company will also be working with them on future submissions, and the FDA reviewers usually have a good memory.

Summary

To a handful of individuals, the entire regulatory process may seem like fun. But for the rest of us, the regulatory process may be about as fun as a root canal without anesthesia. Sometimes the process can be painful because one cannot control biological outcomes from a clinical study, and it is impossible to completely gauge how the FDA will view your data and its sufficiency for approval. Therefore, begin with a sound strategy developed with the help of veteran regulatory experts because this phase is of such critical importance to your company's success. Also be aware that it would be a good idea to always assume that more supporting data will be needed than planned.

Regulatory review timetables are uncertain. The regulatory phase timeframe can be highly variable for both therapeutics and medical devices, even within each of these categories. Plan for longer review times than one would assume and do the best to anticipate the types of additional data and information the FDA may want to see from clinical studies.

Securing regulatory approval is the final milestone in order for a product to reach commercialization, and it is the culmination of all a company's work. Approach the regulatory phase with a well-thought out strategic plan and be sure to execute it well – it will maximize your efforts and minimize risks. Every company wants to complete their clinical testing and regulatory review in the shortest time possible, but do not cut corners and hope to get by with the bare minimum. As the saying goes, "there never seems to be enough time to do it right – but one always seem to find time to do it over."

Chapter 14
Company Life Stages and Changing Management Styles

All companies transition through different corporate life stages. Nascent organizations grow into mature organizations, and this transition occurs subtly over time. Throughout the transition process, business practices evolve. Changes occur in the process of how a company makes its decisions. A mature organization makes decisions differently than a start-up organization, and the optimal process for an early stage company is not optimal for a mature company and vise versa. As companies pass through different life stages, they exemplify particular qualities unique to that stage. Each corporate life stage has a parallel to a similiar growth stage in our own human development. Surprisingly, some of the same advice for human growth stages can be applied to parallel life stages of an organization. Being aware of company life stages is vital, because fundamental problems can occur when an entrepreneurial company transitions to the next growth stage, but their leaders do not. As an organization makes a transition, the biotech leader must also make a transition, particularly in their management and communication style if they hope to be effective in leading the company to accomplish their goals.

Why Understand Company Development Life Stages?

The purpose of this chapter is not to psychoanalyze an organization, nor is it to interpret each organizational change for a parallel to human development. The purpose is to help the entrepreneur be a more effective manager of a highly dynamic business that is dependent on effectively working with highly intelligent and motivated individuals. Rest assured, a company will transition through each life stage without effort on anyone's part. Transition progressively occurs as staff is added, and the company advances their products through the development pathway. Entrepreneur leaders must be certain to adjust their management style during these transition periods, or they may become a source of problems by impeding company progress.

C.D. Shimasaki, *The Business of Bioscience: What Goes into Making a Biotechnology Product*, DOI 10.1007/978-1-4419-0064-7_14,
© 2009 American Association of Pharmaceutical Scientists

Much can be written on corporate life stages, and it is certain that one can develop an entire science of analysis for corporate development and life stages. However, this is not my intent, nor should it be a driving interest of the biotech entrepreneur. It is not essential to become an expert at recognizing each and every development stage of a company; in this context, it is only important to understand when it occurs and when to make proper adjustments. All organizations are dynamic, and the entrepreneur leader can improve their ability to effectively manage for long-term success by understanding corporate life stages. In the early phase of a company, leadership of an organization is contained in one individual, as it grows, leadership is shared by many key individuals; as it further grows, leadership should be shared by an expanding number of managers and key personnel. Thoughout this discussion, leadership can refer to one or more individuals, but it is always in reference to those with the responsibility and authority to make final decisions.

Organizational Life Stages

The life stages of an organization are outlined below, and these represent a continuum that all organizations transition through during their growth:

	Organizational life stage	Business identifiers	Characteristics
Birth	Infancy	Start-up, spin-offs, new organizations	Newness, excitement
Development phase	Toddler	Some success with early product development	Anticipation, thinks they are invincible, unaware of fears
Development phase	Adolescence	More success in product development and funding	Expanding strength, some successes, growth
Development phase	Teenager	Major advances in product development, facing regulatory and financial challenges, able to adapt but awkwardly	Major challenges, critical decisions, company life decisions
Expansion phase	Young adult	Successful and growing	Market entry and success
Expansion phase	Adult	Venerable and established	Achieving market dominance
Expansion phase	Retirement	Long history with product, nearing end of product life cycle, limited product innovation	Unaware that products are past maturity, business model not working any longer

Below is a discussion of the characteristics of a company at each life stage, along with the changes occurring within the organization, and the impact of the leader's management style. Though all companies go through these life stages, the length of time each company spends at any stage is highly variable depending upon their product and its development time to commercialization.

Birth Phase

Infancy Stage

Just as giving birth is an exciting experience so is starting a new company. There is excitement in new life; there is great potential and a bright future. Birth brings the desire to tell others about the new organization and the entrepreneur cannot seem to stop talking about its future. The sense is excitement, anticipation, and newness. This analogy is similar to that of new parents with endless stories to tell about their newly born infant in tow with reams of photographs – all looking identical.

With birth comes responsibility. When my wife and I became first-time parents, anxiety crept into my mind when we brought our first child home from the hospital. While driving home, I began to think – we do not know how to take care of a baby. Thoughts raced through my mind that our child could die because we certainly have no idea what to do! The responsibility for a new life began to weigh heavy on me and our responsibility for her future. I recall thinking I wish there was some sort of manual that came with new babies – the one that tells you how to take care of them, what to do in each situation – everything seemed uncertain. Surprisingly, our daughter survived, and transitioned to beautiful young woman. By the time we had our second daughter we were a bit calmer and less anxious because we then knew what to expect. We had figured out such life-critical things as where to buy formula, and which brand of diapers do not cause diaper rash. Our second child also survived and turned out to be a beautiful young lady, and in spite of my anxiety about being novices, both our daughters are well-adjusted, productive, and live life with purpose. Care, time, patience, and consistency are key ingredients to nurturing a life and a company to reach their potential.

Start-up companies by definition begin at the most fragile growth stage – infancy. Their primary need is sustenance – to be around tomorrow to do more of what they do today. At this stage, the biotech entrepreneur and small team will be doing almost everything themselves. Invariably, everything seems to be important and it must somehow all get done. This means long hours, nights, and working week-ends – it seems like the "baby" never sleeps. The biotech entrepreneur quickly learns that they need to be a jack-of-all trades. A start-up company comes with multiple dependencies, and many times these are tethered to others such as a licensing institution, or to the goodwill of an incubator facility, or to the favor of a potential shareholder.

During this stage, the entrepreneur will be leading the organization with a command-and-control management style. Command-and-control just refers to all decisions being made by one individual who then directs the work of others. The wrong connotation is a military-style of management, where a Sergeant barks orders to everyone, but this should not be the parallel. If the leader is a scientist, this may require directing the experimentation as well as the business matters. Everyone will be looking to the entrepreneur leader for direction to make quality decisions as they will be micro-managing everything. The leader at this stage can neither be tentative, nor continuously vacillate in their decisions. The absence of a strong visionary leader, coupled to poor strategy and poor execution will result in the demise of a start-up before it has a chance to transition to the next growth stage. The leader must know their goals and how he/she will achieve them, even if some are still being formulated along the way.

Development Phase

Toddler

During this life stage, an organization makes a transition to functioning somewhat independently, yet still requires constant monitoring to keep from doing damage to itself. To reach this stage, an initial funding event will have taken place. The organization has boundless energy and possesses the desire and curiosity to chase after many things. This life stage requires a strong leader because if everything seems equally important – nothing will get finished. The leader must constantly reiterate the corporate plans and the priorities to everyone. If the leader is timid, unclear, or inconsistent, the company's growth becomes stunted, and limited resources quickly become depleted – producing no future value. The leader at this stage must still manage with a command-and-control management style to ensure that growth occurs in the right direction, and that all activities are directed toward the same goal. However, there should be team members who are now trusted to carry some portion of these leadership roles. Consistency and persistence are the key elements for companies during this stage of growth. Some organizations never make it past this life stage because of external challenges. The company must keep doing what it set out to do, being persistent, resourceful, and not giving up.

Adolescent

At this stage, the organization appears to be growing, and progress is being made toward its corporate development objectives. Additional funding events have occurred and product development successes are being realized. Employees are happy with the organization and its future; they feel informed about the company and its progress and they enjoy their work. There are beginnings of public interest, and possibly the company may be a future biotech poster child – an example of a promising start-up company.

The leader still manages through a command-and-control style, though there are now others who share some of these responsibilities, and the employees are comfortable with this because it is working well for the organization. Employees know that they can go to the leader for solutions to their problems, and are comfortable with that relationship. The adolescent organization must continue to play to their strengths, and a key strength is its responsiveness and agility. Mature organizations are not capable of making such rapid responses to change because they are not structured in such a manner, and they may not pay as much attention to these external changes.

Teenager

Because of the lengthy product development cycle, many companies spend a long time at this life stage. It also is the stage where large amounts of money will be consumed as companies move toward lengthy clinical trials and regulatory processes. At this stage, the company has secured several significant funding rounds, but now they are encountering product development challenges, and possibly, experiencing regulatory problems. Prior to reaching this stage, a transition in management style must have been in the works. By now the staff and employees have gained sufficient expertise, and possess untapped capabilities, which become underutilized if the leader did not make the transition by this stage. Although the team was previously comfortable with the former management style, they begin to sense a growing frustration, along with an absence of challenge to themselves and their own career. The leadership must have made a shift in responsibility to key managers and employees, who are responsible for the tasks at hand. Leaders at this stage who still direct task-oriented work, rob capable employees of the opportunity to step up and contribute at a greater responsibility level. If the previous management style continues, employees become minimally effective, and find nonproductive things to do that occupy their unused capacity, or worse, they may leave to find more challenging positions of employment. During this stage, the individuals who the leader manages now may possess more information and knowledge about a given situation than the leader does. These employees are the best ones to make these decisions, but because of the previous command-and-control management style, they may be fearful of doing so, or they may think they will not be heard, so they do not contribute at this level. If unchanged, this creates a problem in employee turnover, and can result in good future managers moving on to other companies.

By this stage, the leader should have made the transition to a delegate-and-inspect management style. This does not happen instantly, but gradually, as they let everyone know they will be delegating more responsibility to them, and why. The leader will help themselves and their team, by teaching strategy to them, rather than just approving or disapproving decisions. This is as much a transition for the employees as it is for the leader, who in the past made all the decisions. When the leader is now asked for a decision about an issue, they should explain the important business or scientific goals, the motivation and rationale for the objective, and then

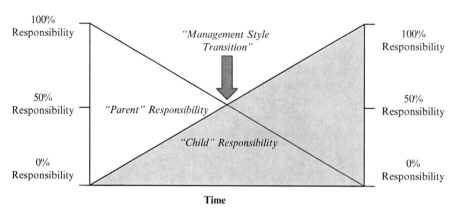

Fig. 1 Management style transition phase and analogy to Parent-Child responsibility transition

ask them for their recommendation based on this strategy. A company should view this transition similar to the shift in responsibility that occurs in a parent–child relationship (Fig. 1).

By this time, the company has grown quickly and the infrastructure has expanded. As organizational structures enlarge, communication channels easily break down. This is also the stage where corporate communication problems arise. Previous frequent and informal communications are now replaced by less frequent and formal communications. The absence of information can cause employees to think and presume things that may or may not be true. Several adjustments can help make a successful transition through this stage. One, is frequent and meaningful communications between the leader, the senior staff, and all levels of the organization about the goals, the plans, and the reasons for the execution choices. This is a key characteristic of successful companies, and this life stage requires lots of communication. Employees also want to know what is going on outside of their own function, and they need this information to help them work more knowledgably. Be sure to communicate the progress in all areas of the business, even if it changes frequently. Work to communicate in as many ways possible. Employees still need to be reminded of the vision and know that there is purpose and value to what they are doing. Also, the leader must show that they live by their core values in their daily activities and in all the decisions they make. Core values that are only framed and hung on a wall, but do not make it into the hearts and minds of the leadership will ultimately be resisted at all levels below management. However, values that are reinforced and shown to apply to senior management will be embraced, and even expected of all staff throughout the organization.

For companies with shorter product development times such as for diagnostics and medical devices, they will spend less time at this stage compared to their counterpart colleagues in the therapeutic field. At this stage, biotechnology organizations have not yet matured. Though they are most likely operating in a relatively new technology area, the tentative nature of success is felt. Like a teenager, there are moments of greatness accompanied by moments of despondency as to the future and their potential. Make a management transition here (technically and operationally), this will improve the chances of moving to the next stage of growth.

Expansion Phase

Young Adult

Organizations that reach this stage are rising stars in the biotech industry. They have successfully passed the last stage of their product development, and they can see a bright future. They have honed expertise, and have recently received FDA approval for their new product, and they are working on building a market. There is tremendous potential in the company, and they are adding staff and expanding their capabilities; they may make a few mistakes along the way, but these are corrected quickly. At this stage, companies must continue to execute their plan well and monitor their external environment for changes that may impact them and their products. Companies need to respond quickly as they did when they were younger and not set up a bureaucracy that impedes their agility and capability to quickly respond. The infrastructure challenge will be coming into balance with the creation of a standard process, but be sure to maintain the flexibility and creativity of the previous development stage.

At this point, the management style has become more similar to larger corporate entities yet the company should still have the ability to manage in a way that takes advantage of a smaller company's strengths. The leadership and management are becoming quite capable of handling development and growth needs, but they still experience a few crises every now and then. During this stage, the focus should be on training more leaders at new levels of the organization, rather than just better managing tasks because these new leaders will be impacting a greater portion of the organization. Communication is still vital and should be a hallmark of the company at this stage.

Adult

Only the most successful biotech companies reach this growth stage. They have fully developed capabilities and are strong producers, manufactures, marketers of their products, and command a strong and growing market share. These organizations are well-managed, and all portions of the business know the organizations, goals and what their departmental functions entail. Adult companies can be thought of as profitable, growing, and forging new territory in the industry. Companies like Amgen, Genentech, Biogen-IDEC, and Genzyme occupy this stage of growth. Yes, the have their challenges, and some of the early employees have left to go on and start companies of their own. Some who were there from the early days remember a different company because by its very nature the company at this stage has become "institutionalized." There are bureaucratic processes that have been adopted, but these have been minimized because there remain enough individuals who remember the importance of just getting the job done, rather than creating more processes.

Retirement

Although this is generally an anticipated time for individuals, for a biotech organization this stage is one to be avoided. Reaching this stage means that either the company has become irrelevant to the market, or that their products have outlived their unique value. Reaching this stage usually means that the company stopped innovating and relied on their same product, which eventually ran out their life cycle and became less relevant in the face of competition. It is less common in the biotech industry for companies to reach this stage, but it still happens when products become commodities, and companies fail to continue product innovation but instead, focus only on sales and marketing. We will not dwell on this stage, but it is management's responsibility to lead and guide their organization through the previous life stages without reaching "early retirement." Biotech companies are known for innovation and product leadership, and as such, it would be unusual to find many biotech companies that reach this life stage in the near future. However, as we begin to see the emergence of "generic" biologic drugs, more companies may succumb to this stage if they do not continue to innovate.

Other Corporate Identity Features

There are many other parallels that can be drawn between organizations and the human condition. Some of these characteristics are interesting but not relevant to the biotech entrepreneur. There are parallels that one can identify in most any personal characteristic; however, we will refrain from elaborating on many of these only to say that a few interesting parallels are listed in the table below.

Individual Personal Characteristics	Equivalent Organizational Characteristic
Personality	Corporate Culture
Individual Expertise	Core Competency
Physical Stature and Strength	Market Dominance
Age	Stage of Growth Cycle
Cultural Background or Ethnicity	Type of Industry

The CEO's Role in the Transition of Life Stages

It is incumbent upon the leadership, especially the CEO, to lead change and to shepherd their company and employees successfully through these growth stages. The CEO must recognize and respond by managing in a life-stage appropriate manner. The management style optimal for a start-up company having only one part-time administrative assistant, a scientist and a technician, is not effective when they reach 25 employees, having a Chief Scientific Officer, Chief Financial Officer, and a Chief Medical Officer. The CEO must change and transition as the organization transitions, just as parents must change how they nurture and relate to their children as they grow.

It is possible that the CEO may have difficulty adapting to changes and may still want to manage as in a previous life stage rather than make this transition. Should this happen, good senior managers will leave, and the organization will develop a decision-making bottleneck at the CEO, and fewer things will accomplished because no one is empowered to make a decision. Encountering resistance to change is not uncommon when an entrepreneur and the start-up organization grows to the next level, or receives a significant round of funding. Sometimes the CEO does not know how to change or they do not want to change. If the company is venture capital-backed, the venture capitalists will have limited patience (rightly so), and will give the entrepreneur an opportunity to right the ship. If they cannot accomplish change, the VCs will replace the CEO and hire a seasoned CEO to replace them. Venture capitalists may be seen as only protecting their investment, but if they are seasoned and experienced, they will quickly recognize the signs and symptoms of an unhealthy organization, and correct the situation such that the organization as a whole can survive. Entrepreneurs must realize that there are many significant challenges to building a successful biotech company, but creating artificial management problems is counterproductive and can be prematurely fatal to the organization.

Work-Life Balance

The work of a biotech entrepreneur is never done. By definition, an entrepreneur always has more work to do than there is time to finish. Throughout this book we have discussed ways to plan, make better use of your time, instill urgency in your team and suggested ways to "jump-start" things, so you can minimize challenges later. Truly, the biotech entrepreneur's life is full and busy but without drive and persistence most of their aspirations will not materialize. However, I must not fail to remind you to be sure and maintain balance in your life even in the midst of exciting yet time-consuming activities of a growing organization. Learn how and when to take a break and spend quality time with family and friends, and know when to just get away for a few days. The success of building a biotechnology company will be hollow if you accomplish this, but lose touch with your family and friends, or you burn out before you see the fruit of your labor. Always remember that no one on their death bed said, "I wish I spent more time at the office". None of us can forget 9/11. The employees in those buildings did not run back into the Twin Towers to retrieve work, files or belongings – they returned for family and friends. Learn to look at your life from its last chapter backwards. Read the end of your story first and it will help give meaning to the time in between. This admonition may seem premature for those that are just starting their budding organization, however, weeks turn into months, and months turn into years – and years become decades very quickly. Refer back to this section in the future, as you will find that you need to be reminded of these words sometime during your long, long tenure as an entrepreneur.

Summary

The view presented here is an application of human life stages during organizational growth. Knowing an organization's life stage will help the biotech leader to manage and lead in synchrony with their growth needs. Like human beings, all organizations go through life stages. Having a fundamental understanding of the life stages of an organization is critical to applying the right knowledge in the right situation and at the stage. A benefit of knowing an organization's life stage is the ability to properly apply other business theory and strategy when applicable. Adopting *any* growth strategy without regard to one's organizational life stage is detrimental because an organization may not be developed enough to properly execute the strategy as intended. Although the transition through corporate life stages happens in a time-dependent manner, the maturation process does not. Just as with humans, though a toddler can grow to an adolescent, then to a youth with the passage of a certain amount of time, it does not assure maturation. I am sure you can think of examples of chronologically "mature adults" who are really just large toddlers. Maturation through corporate life stages is not solely time-dependent, but it is facilitated by the willingness and ability to make appropriate changes at the right time.

The two most critical life phases for a biotech company are the Development Phase and Expansion Phase, and it is important that changes in management style are incorporated during the development phase. Failure to make management style changes are detrimental to company growth, and it impacts the success of the organization. Within the development phase, the company requires two totally different styles of management. Some call this the "entrepreneurial management style" vs. the "professional management style," or the command-and-control management style vs. the delegate-and-inspect management style. Entrepreneurial management is characterized by centralized decision-making with strong control, whereas professional management style is characterized by delegation of decision-making responsibilities and formal control mechanisms. Whatever the style is called, it is important that distinctions are recognized and that changes are made. A management style that worked two years ago may not be appropriate for the organization's growth stage today. Learn to recognize the changing life stages, and the changing leadership needs of an organization.

Chapter 15
When Do You Call it Quits?

A book on the business of bioscience would not be complete without talking about the dissolution of a company. Like it or not, the vast majority of biotech companies do not make it through to commercialization of their products. According to the Small Business Administration (SBA), the success rates for all small businesses in the US is estimated to be 44% during the first four years, though some estimates say only 20% survive after five years. With the additional challenges to business development facing the biotech entrepreneur, it is hard to imagine that the biotech business success rate would be higher than this statistic. Whatever the real success rate – it is painfully low. There are various reasons for company failure, and some of the reasons include the following:

- Poor market assessment before licensing the technology, such that there was no real market interest in the future product and the company could not raise enough funds to complete product development.
- Investor interest in the technology field waned, or was viewed as too early or too challenging for an investment at this time, and the company ran out of money.
- The management team was not capable in the eyes of investors and could not secure the funding needed to continue product development.
- Poor execution of the product development plan, such that the costs escalated far above a profitable return-on-investment for investors.
- The technology failed to produce a viable product during development due to insurmountable technical or scientific challenges.
- The product did not satisfy the requirements for regulatory approval, or the clinical trial results were disappointing and there was no interest in starting new clinical studies.
- A downturn in the financial market and the company did not have enough cash to sustain itself until it could secure the next round of funding.

For the biotech entrepreneur there are two types of conclusions. The first is when the entrepreneur does not possess the advanced abilities, and is no longer needed beyond a particular development stage. The second, is when the company itself is no longer sustainable for the reasons cited above. Both situations are difficult and painful because the entrepreneur and the team have devoted a good portion of their lives and energies to something that they believed in.

C.D. Shimasaki, *The Business of Bioscience: What Goes into Making a Biotechnology Product*, DOI 10.1007/978-1-4419-0064-7_15,
© 2009 American Association of Pharmaceutical Scientists

How Can One Tell When It is Time?

Though there are more failed biotechnology companies than there are successes, it is also unknown what percentage could have reached success had they found the proper funding or had an alternate plan. Nearly all successful companies can tell of near death stories for their business. Michael Dell, Bill Gates, and a host of others who have started successful companies have similar stories. As discussed in Chapter 2, the company is guaranteed to encounter difficult situations and seemingly insurmountable challenges along the way. Realize that many notably successful companies in other industries were turned down multiple times by some of the most reputable financing institutions, and even shunned by business professionals. Take the idea for overnight delivery of business packages and letters that resulted in FedEx. This idea was deemed unworkable and unlikely to be successful from the beginning. One notable venture capital organization had enough honesty and candor to keep track and post the companies that it turned down for funding, which ultimately became successful – including FedEx. These can be found on Bessemer Venture Partners' website under their "anti-portfolio.[1]

Reaching the conclusion that the company is no longer sustainable is not the same as facing typical challenges in raising money, or keeping the business going when it seems no one understands. Nor is it a situation where the company encounters difficult times and the entrepreneur feels like quitting. Also, this is not the same as encountering challenges in the market where getting the right distribution model or target audience is slightly off. Many of us have been at these junctures, did not give up hope, and found that around the corner was the right individual, the right investor, the right contact or needed break into the market. Concluding that the company is no longer sustainable is the point where it is absolutely certain that the company has missed the market and there is no interest in the product, or that the technical challenges are scientifically insurmountable and the company cannot attract any more funding. This is when money cannot be raised no matter what is done to change the story or market approach, and the company has tried every possible way, and there are no more options. Reaching the conclusion to close a company is determined by external factors, but the conclusion for the entrepreneur to leave can be motivated by both internal and external factors.

External Indicators

I can remember being in a "near closing" situation twice in my last position, where no new investors had committed, and there were 25 employees on the payroll with only two months of cash in the bank. I remember hearing an "experienced fund manager" tell me to layoff the research department, pair down the staff to a small group,

---

[1] http://www.bvp.com/Portfolio/AntiPortfolio.aspx

and hibernate until more funds can be raised. We persisted however, and found the additional capital, and moved the product in to commercialization. Sometimes downsizing may be good advice, but this should be done only when there are no other alternatives. Also, be aware of who is giving advice, since everyone has an opinion. Be careful of taking advice from those who never ran or operated an organization, who never had to struggle with making payroll, and whose only advice is to always cut deeply into the core strengths. Though these advisors can do the math very well, they may not understand the effects on the day-to-day functions with their approach, or the likelihood of resurrecting anything from these ashes. Even when all things look hopeless it does not necessarily mean it is time to quit, because sometimes another perspective is needed, or there is a need to seek creative solutions from others.

There are times when an organization will have no other choice then to downsize in order to have a chance at survival. The external indicators that a company has no future and must wind down (rather than downsize and keep going) can sometimes be pretty clear and unambiguous. Therapeutic companies may reach this point when the data from their Phase II or Phase III studies are disappointing, and their product did not prove efficacious, and they have no other products far enough along in their pipeline. Medical device and diagnostic companies may reach this point when their product is in the market but they cannot generate meaningful revenue to sustain the organization, or they cannot attract any partners or financial supporters. In these situations, the fate of the remaining assets and any value left in the technology is usually decided by the Board and major shareholders.

Internal Indicators

Entrepreneurs and CEOs of start-up organization are by nature optimistic, tenacious, and unconventional – which is how they survive the myriad of challenges they face on a daily basis. One internal indicator to recognize is when they transition from optimism, to pessimism and cynicism. It is when the entrepreneur who always saw the "glass half full" now sees nothing in the glass. It is when family members tell the entrepreneur that they have changed, and they are rarely ever happy and seem to find no joy in things that used to interest them. It is when the entrepreneur can no longer see the future as brighter than the present and they no longer voice an optimistic tone – this may be a sign that it is time for change. There are health consequences also, I have seen youthful individuals go completely grey-haired, and individuals who were athletic and healthy, no longer have any energy and sink into apparent depression. There are personnel consequences when one begins to ostracize their family and those who are important to them. At this point, the entrepreneur may need to recognize that they have fought valiantly, and it may be time for someone else to take up the charge, and move on. This internal gauge is a good indicator for entrepreneurs who by nature are never discouraged about the future because to them the future will always be better then the present or the past. The entrepreneur's decision to leave the company may be irrespective of a decision to close the company. However, their departure can also be forced due to their inability to contribute to the organization and its current needs (discussed in Chapter 14 and below).

Assuming There Are No Alternatives, What Should be Done?

A decision to wind-down the company presumes that everything possible has already been done. If funding is the issue, this includes seeking help from the Board members and their contacts in finding capital for the company and putting the company up for sale to a strategic acquirer. The company should have contacted all existing shareholders and informed them of the dire situation, and requested additional financial support equal to their pro-rata share of ownership. If the company has venture capital funding the VCs, they will be heavily involved in this decision and will provide support and options as long as the organization has value; but even they will not put good money in after bad. Any decision to close down the company must be reached with the help of its Board members and major shareholders.

Face Responsibility: Do Not Cover Up, Make Excuses, Deny, or Ignore

A family member relayed to me a situation where an internet start-up company had financial difficuties during the early phase of its development. Over time they noticed that the CEO was not showing up very often for work, and over time his presence became more and more rare. The employees knew that the company had financial difficulties, but they were diligently working to move the company forward in spite of the growing absence of the leader who told them that a financial deal was eminent. The CEO sheepishly avoided talking much about the financial situation and as it turned out, the effort to finance the company through a merger ultimately failed. That day, the CEO called from his home to say there was no more money, and sheepishly asked "are they all mad?" The response was, "No, everyone just wants to know what is going on and get an explanation!" He replied "it was too hard and I couldn't face everyone." Do not run from this type of situation. Face responsibility and deal with it no matter how difficult.

Honesty and Transparency

It is critical to be honest then transparent with everyone, especially your employees. Most individuals are capable of making their own decisions when given the correct information. Unfortunately, during the tough times employees may believe that the management is not being honest or that they are hiding things, and therefore employees may assume the worst, then their actions will be based upon fear rather than knowledge. I always tell every new hire that there are risks associated with working for a start-up company, but they can be sure that they will know how the company is doing on a consistent basis. I also tell them, though there may appear to be security in a larger organization, it may be that they just will not know what is going on until it is too late.

Employee Assistance

If you must wind down the company, you should offer employees access to company resources such as computers, printers, and telephones while they search for other positions during their remaining time under employment. If you are large enough to have personnel services, these resources should be offered to help outplacement of employees find new jobs. Ideally, there should be the ability to provide some severance, though this may be difficult to do. Sadly, many times when companies fail, it is because they ran out of options for funding to meet payroll, so having the option of severance is not always available. If you have to let employees go, do everything you can for your employees and let them know how much they are appreciated.

When It Is Time for the Entrepreneur to Leave?

Sometimes the entrepreneur, founder, or CEO is asked to leave the company. This event should never be one taken by surprise; however, it almost always is. When a company reaches this point, it is usually because the CEO or biotech leader was not paying attention to the signs that the Board or investors were unhappy with the progress of the company, or that they told the CEO things needed to change but the CEO was ignoring them. There may have also been a sense of immutability in this position, and a thought that the company could never survive without them, or that they were too valuable to be fired. These thoughts should never be entertained. If the opportunity for change and restitution has passed, the best thing for the entrepreneur and their future career is to move on to another endeavor. They should then work out the best arrangements possible for support and participation in the upside should the company be ultimately successful. If the decision is clear that it is time to leave and let someone else take over, the CEO will experience many things.

What Can be expected?

The feelings that some entrepreneurs may go through during these times might include denial, regret, anger, bitterness, resentment or sadness.

It may seem normal for someone to experience these emotions, but they should never dwell on them nor let them take hold of their thoughts and life. These are nonproductive and destructive emotions, and they only serve to tie the person down and keep them from having any chance of future success. My advice is to reject these sentiments completely, embrace forgiveness, and keep them out of your life. The entrepreneur needs to be able to move forward, learn from their mistakes and past experiences, then focus their optimistic and constructive ability on another endeavor.

Summary

This business of bioscience creates a business of uncertainty. A company may begin with great ideas but sometimes these lead to failed products. Oftentimes these ideas were truly great but the management could not make it work, or the clinical trials just failed. Whatever the reason, a biotechnology business may reach the point where it is no longer possible to continue. When circumstances become uncertain a leader's natural response may be to withhold information from others rather than to share it. Yet at these times there is a greater need for more frequent communication with employees about what is happening, and what the management is doing to effect change. Employees need to hear from the management about the plans and the progress, and then they can make up their own minds about what they should do. The entrepreneur does not control external events, but they do control information flow. Make a point to have open communications with all your team members. You may be surprised to learn that there may be things they can do collectively to assist with solutions because communication gives employees the ability to participate. Work together with employees and team members to reach solutions that may help improve the position of the company. Ultimately, you need to handle this situation in the best possible manner and do the best to take care of the people who joined in helping to reach these goals. The lessons learned during this time are the hardest but can be some of the most valuable ones for you and your future success.

Chapter 16
Conclusions and Serendipity

Although the business of biotechnology sounds calculated and well-planned, one can always find an element of serendipity buried deep within its roots. Things do not always arise from what was originally intended. Throughout this book the importance of planning, setting goals, and anticipating problems have been emphasized. Success is never possible without good planning – but the truth is, no matter how carefully laid the plans, events do not always unfold as expected. New challenges constantly present themselves, and one needs to know when to press forward, and when to adjust and take advantage of an opportunity for change. Sometimes these "new opportunities" do not first appear like opportunities. Louis Pasteur, the French Chemist and Microbiologist, said "Chance favors the prepared mind." Indeed, when one prepares themselves and is observant, they can take advantage of the opportunity chance provides. Peter Morgan Kash, in his book titled "Make Your Own Luck – Success Tactics You Won't Learn in Business School," describes the importance of constantly being aware of "opportunities" in everyday situations that we encounter but rarely take advantage of. He sites his own example of a financial news network executive dialing a wrong number, and during their conversation he told the executive if he ever needed a person knowledgeable in the Japanese economy and biotechnology to let him know. Six months later, that wrong number turned into a future opportunity to host a TV program for two years. He also described a chance encounter in an elevator with a relative of a head of state, that resulted in diplomatic relations and business opportunities for him in another country. He calls these encounters the "web of life," but they can be also called "missed opportunities," depending upon how we respond. Call it serendipity, opportunity, or divine intervention – what matters most is one's response to each situation.

Recognizing opportunity is the first step. Creatively adjusting to change is the next. Significant events can provide an opportunity to position a product differently if you recognize them. For instance, a diagnostic test for influenza can also become an internationally stockpiled biosecurity tool if it can be repositioned as a rapid screen for the Avian Flu or the Swine Flu. The biotech leader must always be aware of their environment and the opportunities that present themselves. They must be open to course adjustments or changes different from their original plans if it provides a better opportunity of success of the organization. I am not referring to taking short-cuts that reduce quality or compromise results. This is about positioning the company differently, making product development modifications, repositioning products, changing the market segment, aligning the product with a complementary product, partnering with a nontraditional, but successful organization, or building a business differently than traditional.

Strong-willed entrepreneurs believe that through effort and persistence they can impact everything in their environment, and by doing so they believe they will be successful. This level of confidence is an asset to a biotech entrepreneur; however, they will be more effective if they realize that there are things they truly cannot control or affect. Do not interpret this to mean that when problems are encountered entrepreneurs should learn to sing a Doris Day rendition of "que sera sera". Rather, instead of battling every opposition, sometimes the entrepreneur must learn to leverage the momentum rather than always working to change their environment. As a teenager, I enrolled in judo lessons. During practice it was amazing to watch how a small statured "sensei" or teacher, could almost effortlessly toss a much larger opponent across the room at will. The secret was not to counteract, but to leverage the opponent's momentum; and by slightly altering it, one can move them by their own force into the desired direction. Pay attention to the direction and movement of events in your field, these can also be leveraged to a company's benefit.

A Biotech Company Opportunity

I can relate such an opportunity of my own. While walking down the corridor in our Research Park Building, I had a chance encounter with a familiar acquaintance, a scientist who was the Chief Scientific Officer working for a spin-off company from the local medical research foundation. This person was an accomplished molecular biologist, and one who had previously worked in another life science company that was recently acquired. Noticing this large statured man, who usually was always optimistic, having his head hung down and walking gloomily in my direction, I asked him "how is it going?" He replied in a depressed tone, "We are going to shut down the company, because we can't raise any more money to keep it going." He went on to say, "the Foundation has been loaning us money to keep operations going over the past several months, but they are unable to continue. We are going to have our last Board meeting and return the assets back to the Foundation." I was surprised because I was peripherally familiar with their efforts and the technology, which seemed to have scientific merit. I quickly replied in amazement, "Before you do that, tell me about the situation." We then spent about 45 minutes discussing the situation, and I quizzed him on the science and technology, and about their failed attempts to fund the company. Upon ending this impromptu hallway meeting, I asked him to send me any information he had including old business plans and patent information. I told him "don't do anything just yet, but let me look at the information and see if there is anything that I might be able to do to help." The opportunity was there. With very little time to seek capital,

I knew the first thing to do was to revise the business strategy, put together a well written business plan with a clear focus on the market and the investment return opportunity. Within three months we secured an angel investor commitment for $1.25 million dollars, and went on to develop the technology securing a total of $18 million over the next several rounds of funding.

It Was the Best of Times, and It Was the Worst of Times

The French Revolution may seem like a battle similar to building a biotechnology company, and the sentiment of Charles Dickens sums up how the biotech entrepreneur may feel one day to the next. Entrepreneurs will experience both challenges and victories each and every day – prepare for both. Do not be consumed by the challenges, and do not dwell on them for long. Examine the business strategy in light of these challenges – adjust if necessary – and move on to execute it. Problems get magnified the longer we focus on them. This is the reason why the entrepreneur must keep their focus on the goal, which is to establish and build a successful biotechnology company. As we have already discussed, there are a myriad of components that are necessary to achieving this goal.

Every Success Was First Met with Challenges

Almost every successful company can recount stories of how they almost did not make it. These histories remind us that it is not unusual to be surrounded by people who do not believe you can accomplish what you are planning. Successful entrepreneurs can remember the many individuals who turned down support for them along the way. In 1997, the intellectual property for an innovative search engine concept sat at the Stanford Office of Technology Development (OTD), with no takers and no interest from a partner for the license. The Stanford OTD tried valiantly to out-license the opportunity, but no one apparently wanted it. The inventor Larry Page, and his partner Sergey Brin, believed in the concept so much that they said they wanted to take it and start their own company. These partners soon found out they were also turned down many times for financing from successful Venture Capital groups. Finally, they were able to raise seed capital from Andy Bechtolsheim and David Cheriton, who invested $100,000 in 1998. Later they were able to convince venture capital partners at Kleiner Perkins, Caufield and Byers, along with Sequoia Capital, to support their vision in the financial backing of "Google." Adversity and opposition are natural when creating change. Do not think everyone with money will beat a path to your door in the early days. Someone must communicate the vision, and a team comprised of the right people must creatively execute the strategy. In spite of encountering many detours along the way, this team must still head persistently down the path that leads to their ultimate goal.

Summary

The biotech entrepreneur is someone who has a glimpse of the future and works to align their environment toward what they see. Without vision, there is no change, and without great dreams there is no hope. Many devastating medical conditions still plague hundreds of millions of individuals around the world. Biotechnology provides the means to usher significant medical changes for the world's people. However, change will never occur without the vision and perseverance of a biotech entrepreneur. Change requires hard work, and nothing of any value comes easily. Pursue your dreams, and keep one eye on reality – and never lose sight of the vision and goal. Without vision we are destined to the mundane and routine. However, vision alone will not accomplish anything – it takes a careful plan, superb execution, a team of diversely talented individuals, and a leader who inspires them when they do not see how thing can get done.

Dream Big
Inspire Others
Understand People
Align Strengths
Work Hard
Work Smart
Read Voraciously and Listen Intently
Change Your Environment
Adapt
Use Wisdom
Don't Allow Circumstances to Steal Your Dreams
Let Compassion for Others Drive Your Work
Live Life with Purpose

Best Wishes for Your Success!

Chapter 17
Practical Help: How to Get Started?

It may seem odd to have a chapter titled "how to get started" at the *end* of a book. There is a reason for this – most entrepreneurs are eager to get started, however, most entrepreneurs have many started, but as yet unfinished projects at home and at work. Starting a biotechnology company is not an endeavor to undertake without plenty of forethought, guidance, and planning. Often, when there is so much to do, sometimes the most difficult part is knowing what to do first. This brief chapter contains a summary of actionable items to help you get going. These steps provide an action list for those who may not know what to do first in starting a company. To some, these steps may seem remedial, but to others they may provide the needed concrete steps to get the company going.

Starting a biotechnology company is not a prescriptive endeavor. Rest assured, there are established requirements to learn, such as those in the regulatory phase of product development, but during the start-up phase, life is not always conventional. There are many ways that biotech companies initiate and grow, and they vary considerably. A good way to view this process is like a journey. Let us say you live in downtown Manhattan in New York City, and you desire to travel by car to Monterey, CA. There are dozens of potential routes one can take to get there – some well-traveled routes and other less traveled ones. Along the way one may encounter unanticipated detours. However, as long as the traveler (1) knows their ultimate destination, (2) obeys the laws, (3) makes wise use of time, and (4) heads southwest, they will eventually get there. However, someone can complete steps 1, 2, and 3 extremely well, yet if they travel northeast they will never arrive at their desired destination. The successful biotech entrepreneur must do all of these things well.

The remaining information is a brief overview of some practical steps to help jump start the journey down this path. Neither the order nor the tasks are necessary to reach the destination, but they may be helpful to some in getting started.

C.D. Shimasaki, *The Business of Bioscience: What Goes into Making a Biotechnology Product*, DOI 10.1007/978-1-4419-0064-7_17,
© 2009 American Association of Pharmaceutical Scientists

What to Do First

Organize Yourself

The biotech entrepreneur absolutely must be organized because they will be multitasking at levels much greater than ever before. It may seem obvious, but if you do not already have one, get a Day-Timer, Franklin Planner, or an electronic organizer in order to effectively plan tasks during the upcoming days, weeks, and months. There are many organizational helps that come with these planners such as goal-setting methods and tips on getting organized. Learn to pack more into each day, and become more efficient in order to complete the many things that need to get done during each day. Remember, time equals money. More time, equals more costs and more overhead – but money is a precious commodity for a biotech entrepreneur. Resources can be maximized by being efficient with time. Everyone is given the same amount of time each day; the difference is in how we spent. A physician friend once told me that it was not just the smartest that made it through medical training and residency, but those who could also go the longest without sleep and still perform well. Wise use of time is essential to success for the biotech entrepreneur, so learn how to spent wisely. If you know you are not efficient with time, I recommend a helpful book on Time Management from the Harvard Business Review series,[1] alternatively Franklin-Covey has Time Management seminars that also may be of help.

Organize Your Environment

The biotech entrepreneur will need to be organized at a greater level than normal because they will be dealing with volumes of information and managing concurrent activities. One can begin to get organized by picking up several large 3–4 inch binders with dividers, and begin keeping organized notes. This can save time by not repeating something that may have been done a month earlier, such as remembering those who were sent business plans. Divide these sections into at least the following:

- Financing and investor contacts
- Patent issues
- Staffing needs
- Technology issues
- Product development plans
- Company and product name and image
- Business strategy
- Marketing

---

[1] Harvard Business Essentials, 2005. "Time Management: Increase Your Personal Productivity and Effectiveness." Harvard Business School Press, Boston, 150 pp.

- Board and Scientific Advisory
- Contacts
- Miscellaneous

Utilize a Project Management Software

Effective people are able to simultaneously move multiple initiatives in parallel. To do this, you must have instant access to information on all projects, and know the steps that are required to reach these milestones. It is impossible to retain this amount of information in your head, and you will forget more things than you will remember. A great tool to help in planning is a project management software program. If you do not have one you are familiar with, find one that is simple. For those who know how to use Microsoft Project – go for it. If you are not familiar with this software package, I do not recommend this for novices. With MS Project you can plan and build the Empire State Building, but do not spend all your time learning to use a complex productivity tool. There are several other simpler project management tools commercially available. I, particularly, like Mindjet's MindManger as a brainstorming tool, and the companion JCVGantt chart program because these can be coupled together for ease of planning. These programs are simple and intuitive, and provide both an idea capturing tool, coupled to planning software. Whatever project planning tool you choose, make sure that it can be mastered quickly and provides information in a format that you and others can visualize and work from.

Once you have a project management software tool you will utilize it to help manage a large list of objectives such as those below:

- Finding an attorney
- Incorporating the company
- Completing and signing founders agreements and equity plans
- Recruiting your scientific advisors and early board members
- Licensing the technology
- Sourcing vendors and suppliers for a critical supply or service
- Completing testing to obtain proof-of-concept
- Obtaining marketing information about the demand or need for your product
- Filing or defending a patent to issuance
- Negotiating an exclusive license for your technology
- Determining hiring priorities, and hiring this team
- Raising your seed capital
- Finding a corporate partner for your product
- Filing for a federal grant
- Completing a manuscript for publication
- Completing your business plan
- Finalizing your PowerPoint presentation

Begin utilizing the project software to plan out these suggested goals; there are others, but these comprise some of the typical business development and market development goals. However, the largest portion of your planning will be spent on product development goals, and it is vital that you thoroughly and completely detail these plans (see Chapter 7). A good planning software tool will always let you know where you are in the progress toward each goal, and it shows where help is needed to complete any objective. If you are not familiar with project planning, a good book to read is, "Managing Projects Large and Small: The Fundamental Skills to Deliver on Budget and on Time."[2] The steps that follow simply elaborate more on some of the more important objectives you need to complete in order to establish your biotechnology company.

Next Steps

1. **Perform a preliminary market assessment, and identify your business model (Chapters 3, 4, and 7).** Learn all you can about who the ideal customer is for your product. Define your target market. Who are you trying to reach? Who are the Innovators and Early Adopters of your product or technology? Define your Market segment and their characteristics. Who could be good partners for your product development or marketing? How will you distribute your product?

2. **Secure the intellectual property (Chapter 3).** This usually requires a license from an academic institution. Be sure to check out the scientific credibility of the technology and that it lines up with a market that has a significant medical need for your product.

3. **Map out a product development pathway and outline the market strategy (Chapters 6 and 7).** Detail the steps for your product development plan along with the milestones, estimated timeframe and estimated costs. Finalize your market strategy with assistance in market research and consultation from market experts and medical opinion leaders.

4. **Create the business plan and executive summary (detailed in Chapter 9).** You cannot lead anyone or share your vision if you do not have a written plan. Begin by making it a high priority to get on paper a draft version of your business plan. This is the document you will be depending upon most often to gain investor's interest in your venture. Be sure it is succinct and is a "selling" document, not an "educating" document. Do not forget to put a date on the business plan as you will find this is a living document and should be frequently updated. Do not fret about getting this perfected, just get it down on paper and work to improve the weak parts later. This is one area where most entrepreneurs procrastinate and consequently rush to put one together when they have interested investors.

---

[2] Harvard Business Essentials, 2003. "Managing Projects Large and Small: The Fundamental Skills to Deliver on Budget and on Time." Harvard Business School Press, Boston 192 pp.

Plan ahead and start this project early. Detail your returns and exit strategy, show that the market exists, and that you can capture it. Do not forget to discuss the regulatory route and your plans for how to manage it.

5. **Find a good attorney who has start-up experience (Chapter 5).** Your corporate attorney may become your new best friend, as you may need their help almost on a daily basis when starting your company. Find someone with significant experience in biotech start-ups and one who works well with you.

6. **Incorporate the company and get the formal structure secured (Chapter 5).** Establish your organization, select your company name, and create the right entity for securing investments.

7. **Determine your key staffing needs (Chapter 11).** You will want to be operating as a virtual company as long as possible until you secure enough money that will support operational overhead for at least 18 months. This means the higher your overhead, the fewer months you can operate with the same amount of money. Identify the critical goals that will improve your chances of raising the next round of capital. Be sure to get your founders, Board, and Scientific Advisory Board to execute the necessary documents as you establish your company.

8. **Determine the capital you will need to reach your next milestone, and raise it quickly (Chapters 8 and 9).** This is sometimes the biggest challenge for start-up companies. Learn from each investor presentation you give and adjust the next to overcome the criticisms you have heard. Be persistent, creative, and pursue all financial options that present themselves. Do not overlook friends of any of your existing investors.

9. **Work on raising the next round of capital long before it is actually needed (Chapters 8 and 9).** The key takeaway here is to plan your tasks and objectives well, and base them on your business model strategy. Do not forget about government grants and local money from municipalities, and government programs for life science initiatives.

There is nothing prescriptive about the steps discussed here, just be sure to have a thorough plan and execute it well. As you get started be sure to also seek the help of good resources such as, local business schools, successful biotech entrepreneurs, and governmental business development organizations that specialize in life science support. Starting a biotechnology company is exhilarating, challenging, time consuming, and rewarding. Plan your time and tasks well, and utilize as many helpful tools as possible. Always be sure you are always headed in the right direction and not just following the path of least resistance. Keep moving forward until you reach your ultimate goal.

Glossary

Some of these terms may have broader definitions and more than one meaning in different contexts, in those cases, these terms are defined as they pertain to usage in the realm encountered by a biotech entrepreneur.

Funding and Financing

Angel Investor
High net-worth individuals who are qualified as Accredited Investors as defined by the Securities and Exchange Commission (SEC). Angel investors usually invest at early stages in a company such as the Pre-Seed and Seed Capital Stage.

Acquisition
The process of one company taking over another company in exchange for something of value. This is usually an event that allows existing investors to exit their investment and the entrepreneurial company to obtain resources from a larger organization.

Antidilution Provision
A method used to protect investors in the event that the company subsequently issues equity at a lower valuation than the previous round. There are various arithmetic methods (e.g., full-ratchet, weighted average) by which this is accomplished, but the end result is that these investors receive the benefit of a lower conversion price for their shares.

Balance Sheet
A financial snapshot of the overall condition of the company at a particular point in time. The Balance sheet shows the Assets, Liabilities, and Shareholders' Equity. The reason it is called a balance sheet is that the Assets always equal the Liabilities plus the Shareholders' Equity and they balance out.

Bankruptcy
A legal event when a company becomes insolvent due to its inability to pay its bills or sustain its operations. For companies with no other option, this becomes an event where any remaining assets get distributed among the creditors of the company.

Book Value
A financial value for the company that is calculated from the balance sheet, by subtracting the total liabilities from the company's assets. This is equal to the Shareholders' Equity.

Bridge Loan
These are short terms loans that are used to assist the company in reaching a longer-term financing. These usually have some premiums attached to them because they typically occur at a time when the company may be close to running out of money.

Capital
For general purposes, it is a reference to cash and the use of cash.

Cash Flow
The movement, or pattern of movement, of money into and out of the business. A Cash Flow statement is used to show the cycle of cash inflows and cash outflows into and out of the company.

Closing
This is the date or time at which all the legal documents are signed and the funds are transferred.

Convertible Loan or Note
A feature of a loan or note, that provides the option of conversion of principle and interest into securities, at either the option of the lender or borrower. Often, this is a feature that avoids the issue of pricing or valuation of the company until a larger round of funding is secured, and the specific terms of the funding round are determined.

Conversion Rights
These are the rights under which investors holding preferred shares may convert their shares into common shares.

Controller
The company's chief accountant; also called the Comptroller.

Current Assets
The sum of the cash, cash equivalents, securities, and other assets that can be converted into cash in less than One year. Total Assets include the Current Assets and Long-Term Assets and any other company assets.

Due Diligence
This is the process of thorough evaluation of a company and their operations, and any material issues that would have a bearing on the decision to invest or not to invest in a company. This usually occurs prior to a major investment in a company by Venture Capital or organized Angel Investors.

Debt Financing
Financing that is obtained through the issuance of debt such as loans which are usually secured against company assets.

Depreciation
The reducing of the value of an asset for accounting purposes. This is a noncash accounting event that is based upon the reduction of the useful life of the asset.

Dilution
The reducing of the fractional ownership of the company that is held by investors. Dilution usually affects the previous investors upon a new round of financing, if the previous investors do not add their pro-rata share of new money to maintain their fractional ownership position. This dilution effect may be somewhat mitigated with antidilution provisions.

Dividend (dividend preferences)
These preferences are usually included with issuance of preferred shares. Dividends can be cumulative automatic, or noncumulative nonautomatic. They also make allowances for a small return for the investor should the deal not materialize as expected.

Dry Power
A slang term, used to refer to having more cash available to cover future obligations. Many times it is in reference to the ability of existing investors to fund additional rounds of investment in the company should addition funds be needed for company development.

Exit Strategy
The method of obtaining a return for investors in a company. This can happen via several routes, such as a company acquisition, or an initial public offering where the investors have marketable securities that can be freely traded for liquidation.

Equity
The fractional ownership in a company though common or preferred stock.

Equity Financing
Raising capital for the company by the selling of common or preferred stock in the company.

Fiscal Year
The 12-month period where the company's financial records start and conclude. This does not have to correspond to the calendar year. Some fiscal years begin on July 1 and end on June 30, and others begin on October 1 and end on September 31. Fiscal year choices can be made for seasonal revenue issues or for auditing timing reasons.

Fixed Assets
Assets that are not expected to be converted to cash within 12 months; these include real estate, equipment, furniture, etc., and are sometimes referred to as property, plant, and equipment.

Fixed Costs
These are costs that do not change in proportion to the variation in activity that the company conducts. Examples include rent and utilities, but can also include certain employee salaries.

Gross Profit Margin (GPM)

The difference between the gross revenue minus the cost-of-goods or cost-of-services sold. This is also expressed as a percentage, which is calculated by the gross profit divided by the gross revenue. Gross profit margins for the biotechnology industry can generally be in the range of 50–90%.

Goodwill

This is an intangible financial asset, that shows up on the company's balance sheet when a company acquires another company and pays more than the book value of its assets. The goodwill is attributed to the company's name, its brand and reputation, and value to the acquiring organization. The excess paid is referred to as Goodwill.

Income Statement

A financial statement of the company that includes the income (sales or revenue), expenses, and net profit. Sometimes this statement is also called a Profit and Loss statement.

Indirect Costs

Costs that are incurred, but not directly attributed with the manufacturing or selling of a product. These can include costs-of-services of accounting, and other support services. Opposite of direct costs.

Inflation

The upward movement for prices for goods and services in an economy expressed in annual or monthly percentages, from the same period the previous year. Inflation is usually measured by the Consumer Price Index and the Producer Price Index.

Insolvent

When a company is unable to meet its financial obligations.

Initial Public Offering (IPO) or Public Offering

The first sale of stock of the company to the public. This is a liquidity event for the investors because there is a market for the securities that the investors own.

Investment Bank

A financial intermediary that underwrites the offering of securities of the company during a public offering or private offering. They typically perform these services for a percentage of the proceeds and options to purchase their stock.

Limited Partners

For venture capital, these are the individuals or organizations that provide the capital for the funds which are used to invest in companies. They do not participate in the decision-making of any of the fund activities or investments, and are protected from liabilities through this legal structure; whereas the General Partners are the managers of the fund, and the ones that make the decisions on investments and day-to-day operations of the fund.

Liquidation Preferences
This refers to the number of times investment money is returned to a preferred shareholder in the event of a liquidation of the company, prior to the distribution of proceeds to any other shareholders. Usually this is expressed in a multiple of the investment, but can also be 1×.

Mezzanine Financing
This is a late-stage financing that usually occurs prior to an IPO for the company.

Pitch
The speech or words used to interest potential investors in your company. Sometimes this is referred to as the "Elevator Pitch," because it is meant to be short and catchy enough that you can tell to someone riding up a couple floors in an elevator, and not lose their interest.

Premoney Valuation
The value of the company just prior to the current funding round being considered. One way premoney is assessed is by the number of outstanding and issued shares, multiplied by the last price per share paid.

Postmoney Valuation
The valuation of the company just after the new round of funding, which includes the new money invested. This is assessed by the new number of outstanding and issued shares, multiplied by the current price per share.

Price Protection
Also known as antidilution protection (see antidilution).

Private Placement
This is an investment round that is placed to a select group which is not available to the public at large. It refers to the offering that will be presented to select investors, and has differing Securities and Exchange regulations than those that are offered to the general public.

Pre-Seed Capital or Proof-of-Concept Capital
The earliest stages at which a technology or concept for a product is being proven.

Preferred Shares
Refers to shares that have preferences over any other shares in the company. These can be designated by series such as A, B, C, etc, and have conditions that accompany them that give the investor more control, benefit or rights than others.

Return on Investment (ROI)
This is the profit or loss from an investment. It is usually calculated as an annual percentage rate over the life of an investment. This is a figure that is important in determining the worthiness of undertaking a project.

Redemption Rights
A provision, usually contained with preferred stock, that requires the company to buy back shares from the investor after a specified period of time, for the original purchase price plus a premium.

Registration Rights
A right, entitling the investors to force the company to register shares of common stock, converting their preferred shares so they can be sold such as in an IPO. Unregistered shares in the US cannot be sold.

Rights of First Refusal
A contractual provision that gives the holder a right to purchase something before it is offered to another party; usually made in reference to additional shares of the company in another financing.

Seed Capital
The financing needed after proof-of-concept in a company's development. Occurs at the very early stage of the formation of the company and is relatively a small amount of money, that can range from about $25,000 to about $200,000 dollars.

Start-Up
A company at the earliest stage of formation. This can be a company with no employees and no funding and just a concept, to a handful of employees after initial formation that already has a clear company focus.

Syndicate
The collective joining of more than one capital investment group together to fund a company with each group putting in a portion of the money needed to fund the company. Usually this occurs with institutional investors such as Venture Capital in order to spread the risk and to share the monitoring responsibility of an investment. Generally one VC will be designated as the "lead" investor, and the others will be "coinvestors."

Term Sheet
A nonbinding agreement that sets forth the terms and conditions under which an investment would be made into a company. It usually contains boiler plate language, but is the template for the final legal documents. A Term Sheet is presented to a company at the time at which there is serious interest in the organization.

Underwriter
An investment banking firm that has committed to taking the unsold securities of the firm, which it is offering from its own books. Underwriters are most commonly used at the time of a company's public offering of stock.

Venture Capital
A firm that has raised money for the purposes of investing in companies that are not yet bankable. Venture firms are managed by general partners and they invest other investor's money. The other investors are called limited partners. Venture capital firms usually invest at a later stage than do Angels.

Venture Capitalist
A professional money manager who manages the funds that are used to invest in high risk ventures.

Voting Rights
The right of holders of stock to have a vote in certain company decisions. Various types of voting rights can be granted to holders of preferred shares. These can be in the form of preferred shares voting separate from common share or they can be the only class of shares that vote on certain issues.

Working Capital
The amount of cash or liquid assets that the company has available for general corporate purposes. For more established companies, it is one measure of the company's ability to pay its bills and operate the business. Working Capital is calculated by taking the current assets and subtracting the current liabilities.

Intellectual Property

Composition of Matter
This refers to a specific type of patent where the claims are based upon seeking coverage for the composition of the invention.

Continuation-in-Part
A new patent application that is filled before the earlier patent application or portion of a patent application is abandoned, which adds additional material to the earlier application.

Notice of Allowance
The written communication from the US Patent and Trademark Office (USPTO) stating that the patent is being allowed or awarded.

Patent Search
The formal process of searching for previous patents or Prior Art that can either support or discredit the novelty of a patent being considered.

Prior Art
Existing information or patents that are used to evaluate the patentability of a new patent. Prior Art can be supportive or detrimental to the awarding of a new patent.

Trade Secret
Intellectual property that is held by a company through maintaining secrecy from the public rather than applying for a patent protection. This is preferred when it is more advantageous to keep the information out of the public or when the possibility of receiving a patent for the intellectual property is low.

Technology Transfer
The process of taking a technology out of a university or academic institution and into a private company. There are technology transfer offices now in most Universities and academic institutions that facilitate these licensing activities.

Corporate

Biotechnology
Any application of engineering and technology applied to life sciences. Usually, it refers to applications that use living organisms in the making of its products.

Board of Directors
A group of individuals chosen to govern the affairs of a corporation or business entity usually elected by the shareholders. The Board of Directors holds a fiduciary responsibility for the company.

Critical Path
Within a sequence of activities that are necessary for the completion of a project or to reach a goal, it is the most essential and/or the most lengthy task needed to complete the project.

Downsizing
The reduction of the number of employees in a company through termination, layoffs, or early retirement.

Chairman of the Board
The highest ranking officer on the Board of Directors of a corporation. This individual typically presides over the Board of Directors' meetings.

CEO
Chief Executive Officer, the highest ranking individual in the organization, who is responsible for the final decisions and direction of the organization. The CEO reports to the Board of Directors.

CFO
Chief Financial Officer, the most senior financial executive of the company

COO
Chief Operating Officer

CSO
Chief Scientific Officer

Corporate Structure
The inter-relationship of different groups/departments and their functions that make up the company.

Exit Interview
The interview that is held between an employee who is leaving through resignation or termination, and a company representative.

FDA
US Food and Drug Administration

Good Manufacturing Practices (GMP)
Also referred to as cGMP or "current good manufacturing practices." These are standard guidelines set by the FDA to ensure the quality, reproducibility, and safety of manufacturing processes for manufactured products.

Quality Assurance (QA)
Defines a set of practices which insures confidence that a product is produced consistently and according to specifications. These practices can include inspection, audits, standard operating procedures, and other testing measures.

Quality Control (QC)
The process and policies that are used to insure that the company's products and services are of the highest quality.

Recombinant DNA
Genetically engineered products or processes that combine DNA fragments from two or more sources with the use of restriction enzymes. Also referred to as gene splicing.

Legal

Bylaws
The rules that govern a corporation which include the procedures for holding shareholder meetings, the election of the Board of Directors and Officers, and the issuance of stock in the corporation.

Articles of Incorporation
A legal document filled with the state in which the company is incorporated, stating things such as the purpose, name, amount and type of stock issued.

Certificate of Incorporation
A legal document from the state where the company is incorporated certifying that the company has complied with the requirements of the state for incorporation.

Collaboration
The interaction of two or more organizations working together to achieve a common goal. This arrangement is usually evidenced by a Collaborative Agreement, which states the responsibilities of both parties and the conditions under which the work will be done and how the benefits will be shared.

Confidential Information
Information that is private to the company and its employees. The information is covered by law under Confidentiality Agreements.

Employee Stock Option
The right given to employees to purchase and sell a specific type of company stock at a specific price and time period in the future. These are usually used as noncash incentives by start-up technology companies to recruit and retain needed employees.

Freedom of Information Act (FOI)
The FOI was enacted in 1966 and is a policy pertaining to information that must be made freely available to the public.

Insider
Is a person who by definition owns 10% or more of the corporation or is an officer or director of the corporation. It is also an individual who has inside information about the company and its activities. An Insider has specific legal obligations and penalties regarding the disclosure of Nonpublic information which can be used for trading securities.

Medical Device
This can be broadly defined as any physical item used in medical treatment, diagnosis, or testing. These can range from tongue depressors to cardiac pacemakers. Medical Devices have broader and specific legal definitions outlined by the FDA and have specific requirements that regulate its approval and usage by the public.

Noncompete Agreement
A contract between an individual and a company that prohibits the activities of the individual in certain markets should the individual leave the corporation and join a competitor or begin their own business that is directly competitive to the company. Some consideration must be given to the individual to make this agreement valid and some states do not favor enforcing Noncompete Agreements unless they are narrowly defined.

Power of Attorney
A legal document allowing an individual to act as an agent of another.

Marketing

Brand
The collection of assurances that is associated with a company's products, name, and symbols, which promise the customer a certain expectation of quality or service that has been built over time by their reputation.

Commercialization
The phase when products and services developed by a company are ready to go to the market. It can also refer to the process by which the market is built and expanded in order to accept and purchase the company's product or service.

Competition
In business, it is the rivalry between two or more companies that sell products or services to the same market.

Demand
A measure of the amount of goods and services that a segment of consumers will want to purchase at a given price.

Direct Sales Force
A group of sales individuals that market the company's product directly to a particular segment.

Distributor
A company or group of individuals that specialize in marketing a company's products or services. This organization usually can purchase the products at wholesale and sell them to a retailer.

Distribution Channel
The collection of steps or the route a product takes as it passes from seller to buyer.

Early Adopter
Usually refers to the individual or group of individuals that embrace or purchase new products during the beginning of their introduction to the markets.

Launch
Refers to the beginning of commercialization of a company's products or services.

Market Forecast
The expected or anticipated volume of sales for a product or service prior to its actual sale by examining various available information.

Market Share
The percentage the company captures of the total sales possible for a given product or service in a particular segment.

Market Penetration
The percentage of the market that the company has successfully controlled by its products or services.

Segmentation
A dividing of the market into groups of customers that have similar characteristics or interests to improve product sales by appealing more effectively to homogeneous customers needs.

Additional Reading

Claudio Fernandez Araoz, 2007. "Great People Decisions – Why They Matter So Much, Why They Are So Hard, and How You Can Master Them." Wiley, Hoboken, NJ, 336 pp.

Constance E. Bagley and Craig E. Dauchy, 2003. "The Entrepreneur's Guide to Business Law," 2nd Edition. Thomson, South-Western West, 730 pp.

Cynthia Robbins-Roth, 2000. "From Alchemy to IPO – the Business of Biotechnology." Basic Books, New York, 253 pp.

Dawn Iacobucci, Editor, 2000. "Kellogg on Marketing: The Kellogg Marketing Faculty, Northwestern University." John Wiley & Sons, Hoboken, NJ, 427 pp.

Geoffrey A. Moore, 2006. "Crossing the Chasm: Marketing and Selling Disruptive Products to Mainstream Customers," Revised Edition. Collins Business Essentials, HarperCollins, NY, 227 pp.

Hal Urban, 2003. "Life's Greatest Lessons: 20 Things that Matter," 4th Edition. Fireside Book, New York, 165 pp.

Harvard Business Essentials, 2003. "Managing Projects Large and Small: The Fundamental Skills to Deliver on Budget and on Time." Harvard Business School Press, Boston 192 pp.

Harvard Business Essentials, 2005. "Time Management: Increase Your Personal Productivity and Effectiveness." Harvard Business School Press, Boston, 150 pp.

Hugh B. Wellons and Eileen Smith Ewing, 2007. "Biotechnology and the Law". American Bar Association, New York, 921 pp.

Jim Collins, 2001. "Good to Great: Why Some Companies Make the Leap...and Others Don't," Collins Business, New York, 300 pp.

Louis V. Gerstner, Jr., 2002. "Who Says Elephants Can't Dance? – Inside IBM's Historic Turnaround." Harper Business, New York, 372 pp.

Michael G. Pappas, 2002. "The Biotech Entrepreneur's Glossary," 2nd Edition. M.G. Pappas & Co, Shrewsbury, MA, 184 pp.

Orville C. Walker, Jr., Harper W. Boyd, Jr., John Mullins, Jean-Claude Larreche, 2003. "Marketing Strategy: A Decision-focused Approach." McGraw-Hill Irwin, 4th Edition, Boston, 365 pp.

Peter Morgan Kash with Tom Monte, 2002. "Make Your Own Luck – Success Tactics You Won't Learn in Business School." Prentice-Hall, Englewood Cliffs, NJ, 231 pp.

Steven K. Gold, 2004–2006. "Entrepreneur's Notebook – Practical Advice for Starting a New Business Venture." Learning Ventures Press, 217 pp.
Steven Rogers, 2003. "The Entrepreneur's Guide to Finance and Business – Wealth Creation Techniques for Growing a Business." McGraw Hill, New York, 340 pp.

Index

LaVergne, TN USA
06 October 2009

159944LV00002B/1/P